DESCARTES AND HIS CONTEMPORARIES

DESCARTES
AND HIS
CONTEMPORARIES

MEDITATIONS, OBJECTIONS,
AND REPLIES

EDITED BY

ROGER ARIEW AND
MARJORIE GRENE

THE UNIVERSITY OF CHICAGO PRESS
CHICAGO AND LONDON

Roger Ariew is professor of philosophy and Marjorie Grene is adjunct professor of philosophy and science studies at Virginia Polytechnic Institute and State University.

The University of Chicago Press, Chicago 60637
The University of Chicago Press, Ltd., London
© 1995 by The University of Chicago
All rights reserved. Published 1995
Printed in the United States of America
04 03 02 01 00 99 98 97 96 95   5 4 3 2 1

ISBN (cloth): 0-226-02629-9
ISBN (paper): 0-226-02630-2

Library of Congress Cataloging-in-Publication Data

Descartes and his contemporaries : meditations, objections, and replies / edited by
    Roger Ariew and Marjorie Grene.
        p.   cm.
    Includes bibliographical references and index.
    ISBN 0-226-02629-9. — ISBN 0-226-02630-2 (pbk.)
    1. Descartes, René, 1596–1650—Influence.   2. Philosophy,
    Modern—17th century.   I. Ariew, Roger.   II. Grene, Marjorie Glicksman, 1910–   .
    B1875.D37   1995
    194—dc20   95-6453
                                                                                       CIP

CONTENTS

ABBREVIATIONS

AT   *Oeuvres de Descartes,* edited by Charles Adam and Paul
          Tannery, 12 vols., rev. ed. (Paris: Vrin, 1964–76)
CSM   *The Philosophical Writings of Descartes,* translated by
          J. Cottingham, R. Stoothoff, and D. Murdoch, 2 vols.
          (Cambridge: Cambridge University Press, 1985)

Unless otherwise indicated, all translations in the various essays are by
the authors.

PROLOGUE

M odern philosophy, we have been accustomed to think—or at least to say—begins with René Descartes' *Meditations*. In 1641, in the first edition of that seminal work, Descartes gave us the *cogito:* "[I]t must be acknowledged," he wrote, "that this pronouncement, *I am, I exist,* whenever I assert it or conceive it in my mind, is necessarily true."[1] With this great turning point, suddenly, we were said to have reached a new level of awareness, at which we could ask, not "Scholastic" questions about being, but reflective questions about ourselves: Can we reach out from consciousness to an external world? Who are we as minds in relation to our bodies? and so on. True, the *cogito* had been anticipated in the *Discourse on Method* of 1637, but it was presented in its full glory in the *Meditations* of 1641. Thus this text, in its terse, beautifully wrought six movements, appeared to serve, singly and in isolation, as the turning point from an antiquated, authority-laden metaphysics to a new reflective, epistemologically more self-conscious style.

Such a view may have been comforting; at least it made the philosophical curriculum easy to organize. But by now it has become clear to many that it was badly mistaken. There is tradition as well as innovation in the Cartesian text, as there is in every text. All thinkers start somewhere, within a tradition that has made them what they are, however explicitly they may appear to reject it. Even the most radical innovator has roots; even the most outrageous new beginner belongs to an intellectual community in which opponents have to be refuted and friends won over.

If this is true of philosophical reformers in general, it holds conspicuously for the case of the Cartesian *Meditations,* which were presented before official publication to a select group of scholars, in order that their comments and Descartes' replies could be published in one volume with the *Meditations* themselves, for all who run to read. Thus the edition

1. AT 7:25, ll. 11–13.

of 1641 was not the *Meditations* alone but a compendium: introductory remarks that set the new text in relation to questions already raised about the *Discourse* four years earlier, the six meditations themselves, and then objections from other scholars, together with the author's replies to those objections. To see the *Meditations* more clearly for what they are, therefore, we need to look at the *Objections* and *Replies* and beyond them also at the objectors. We need to try to understand who the people were to whom Descartes was offering his work for criticism, and we need to understand the reasons they had for their appraisals of the Cartesian arguments, as well as Descartes' reasons for replying as he did. Further, in working toward such an understanding, we may find not only where Descartes and his critics fit into the intellectual environment of a tumultuous and fateful time, but also ways in which Descartes' own thought developed as he was articulating his responses. It is the intent of this volume to forward both these aims: to add to our insights into the nature of Descartes' own philosophizing as well as to relocate the *Meditations,* if we can, in the cultural milieu in which they were conceived, written, and presented to the public.

The chief person who managed the circulation of the text of the *Meditations* to most of its first critics was Descartes' friend Marin Mersenne, a member of the strict order of Minims, who from his cell in the Couvent des Minimes at the Place Royale served as the center and informal coordinator of a wide and diverse intellectual circle. From his retreat in the Netherlands, Descartes was in constant correspondence with his monastic friend. However, as Jean-Luc Marion reminds us in his essay, it was the *Meditations* plus the First Objections and Replies that Mersenne received for further circulation. Descartes had first sent the text to two Dutch acquaintances, Bannius and Bloemaert, who turned it over for comment to their colleague, Johan de Kater, or Caterus, as we know him. Subsequently, it was through Mersenne's agency that the remaining five sets of objections were obtained, making six altogether in the first edition; a seventh set followed these in the second edition of 1642. Although it has sometimes been asserted that Mersenne wrote the second set himself, it now seems pretty clear, thanks to Daniel Garber's research, reported in his essay below, that some of the arguments presented in the Second Objections stem from one J.-B. Morin, a professor at the Collège Royal, with whom Descartes also had some correspondence. Officially, however, the objectors may be listed as follows:

1. Caterus, with remarks addressed by him to his friends Bannius and Bloemaert, to be conveyed to Descartes.

2. "Theologians and philosophers," described in the French edition of 1647 as "collected by Mersenne."

3. Thomas Hobbes, described in 1647 as "a famous English philosopher."

4. Antoine Arnauld, whose objections were addressed to Mersenne as intermediary.

5. Pierre Gassendi.

6. A group described in 1647 as "various theologians and philosophers," once more collected by Mersenne.

7. Pierre Bourdin (in the second edition).

It is the *Meditations, Objections, and Replies,* together with Descartes' correspondence and other works by and evidence about the objectors, that form the principal sources for this study. If we hope, through analysis of and research into the background of these sources, to add to our knowledge of Descartes' philosophy and of the cultural context in which it arose, it is Descartes himself who offers us this opportunity, since we are looking at the *Meditations* in the setting in which their author took pains to present them: as one step or series of steps in a continuing conversation with his contemporaries.

We can specify the menu he offers us a little more closely: the First and Seventh Objections exemplify Descartes' relation to the Scholastics of the period and in particular (for the Seventh) to the Jesuits. The Second and Sixth place him directly in Mersenne's circle, that is, in debates characteristic of the developing sciences of the time. The Third and Fifth, submitted by Hobbes and Gassendi, place the Cartesian enterprise in its relation to materialist or Epicurean movements of the day. And the Fourth Objections, by Arnauld, exhibit the profound kinship, and at the same time the differences, between Descartes' relation to the Augustinian tradition and that of the Port-Royal group and Jansenism in general. Thus an intensive study of the *Objections* can further our understanding of the sources of modern philosophy and of the origins of modern science in a manner that is both focused—on a small body of literature and a small number of authors—and wide-ranging—in involving a variety of themes characteristic of the period, from the concerns of Jesuit scholars through a variegated collection of "Scholastic" problems to the deeply Augustinian thought of Port-Royal and on the other side to the more secular themes connected with the revival of Epicurus and other materialist styles of thought. Such studies, we believe, not only contribute to the intensive examination of Cartesian texts but also place them squarely in their contemporary context, illuminating in some detail a number of interacting and contrasting motifs of seventeenth-century thought. Special problems raised include, for example, the nature of Dutch Scholasticism in the period and its relation to Scholasticism elsewhere in Europe, as well as the question of the role of Scholastic thought in Descartes' development. His "war with the Jesuits," as he once called it, and its peaceful solution also demand attention. For the Mersenne circle, there are questions about the contacts of Mersenne with such thinkers as Hobbes or Galileo; in light of his far-ranging interests, the unambiguous role Mersenne has been assigned as public relations man for Descartes seems to need reexamination. In the case of Hobbes and Gassendi, questions arise about the role of materialist thinking in this period, in its relation both to religion and to traditional philosophy. The specific problems dealt with in this rich and extensive field depend in part on the special interests of our authors. We offer not a survey but a range of particular studies and of particular questions worth pursuing as research of this type continues to develop.

This volume brings together essays on the *Objections* and the objectors by a number of scholars whose work exemplifies the attention to context that Descartes' mode of publication in 1641 seems especially to demand. Until recently, most work on Descartes, or on other major philosophical figures, consisted largely of close textual analysis, sometimes juxtaposed with problems of interest to our own contemporaries. A magisterial work like Martial Gueroult's *Descartes According to the Order of Reasons*, for example, consists chiefly of a close examination of the text of the *Meditations* themselves. Edwin Curley's *Descartes against the Skeptics* might be seen as putting Descartes in some sense into context, in stressing his opposition to skepticism, but recent work by some of our contributors as well as others seems to move in a more inclusive (or, if you like, "externalist") direction.

For example, there is Marion's extensive work on Descartes and his Scholastic predecessors. Theo Verbeek has recently published an authoritative work on Descartes' relations with Dutch philosophers and theologians. Jean-Robert Armogathe has written authoritatively on the vexed question of Descartes and the Eucharist. Peter Dear has published on Mersenne in relation to the Schools and also on Jesuit mathematics and experimentation in this period. Garber's essay on the *Meditations* as dialogue in A. O. Rorty's collection of essays on the *Meditations* in a way set the stage for this kind of work, anticipating as it did the general theme of Marion's essay here. Since then, Garber's book on Descartes' physics has shown how fruitful work in this direction can be. Vincent Carraud's recent work on Pascal as well as Thomas Lennon's *Philosophical Commentary* in his translation and edition of Malebranche's *Recherche* also exemplify the historical approach we have in mind. Steven Nadler has published both on Arnauld in relation to Descartes and on Malebranche in relation to both Descartes and the Augustinian tradition.[2]

All these studies, and doubtless many others, exemplify the kind of perspective we are referring to, in which even the most subtle texts, however well worth studying in themselves, are seen in their historical contexts as expressions of the intellectual concerns of their time. We hope that a volume like this one will not only add to the developing literature in this area but also help to stimulate further research of this kind.

One of the many steps we took before producing this volume was to hold a conference to try out some of our ideas.[3] We wish to acknowledge the support our 1992 conference received from the National Endowment for the Humanities, an independent Federal Agency; the French Ministry of Foreign Affairs, and the College of Arts and Sciences and the Research Division of Virginia Tech. We are grateful to all these agencies for their generous assistance in making possible the discussions that ultimately gave rise to this volume.

2. Marion 1981, 1986b; Verbeek 1992; Armogathe 1977; Dear 1987, 1988; Garber 1986, 1992; Carraud 1992; Malebranche 1980, 757–848; Nadler 1989, 1992.

3. This was one of a number of international conferences commemorating the 350th anniversary of the *Meditations, Objections, and Replies.* Other locations included Reading, Paris, and Lecce.

— ᦔ 1 ᦕ —

THE PLACE OF THE *OBJECTIONS* IN THE DEVELOPMENT OF CARTESIAN METAPHYSICS

When we consider simply the corpus of the *Meditationes de prima philosophia,* one question demands our attention: If one cannot refrain from taking the six meditations for the body of the work, should one add the *Objections and Replies* as merely related pieces (*integumenta,* Descartes would have said) or recognize them as inseparable members of the central body? In other words, do we have in the volume of 1641 a short work burdened with a confused apparatus of academic and far from methodical discussions, sometimes tedious and often useless, or an organic whole, in which the *Replies* play an essential, albeit secondary, role in relation to the six meditations?

This question determines in large part the whole interpretation of the *Meditations.* And first for an obvious reason: certain decisive doctrines appear only in the *Replies,* while the *Meditations* overlook them—doctrines such as *causa sui,* the understanding of the divine essence as power, the physical explanation of the Eucharist, the positive indifference of God, etc. It goes without saying that the status of these doctrines will be entirely different depending on whether the *Replies* are taken as a facultative annex to or as an essential consequence of the *Meditations.*

The reply to this question must begin from an obvious fact: from the time he planned to publish his first philosophy, Descartes envisaged joining objections and replies indissolubly to the body of his demonstration itself. Nevertheless, this very fact raises problems: we still need to understand why Descartes intended this joint publication.

The Responsive Structure of the *Meditations* in Relation to the *Discourse on Method*

The first explicit indication of the role of objections in the *Meditations* is found in the Preface to the Reader. There, Descartes specifies that he had asked in the *Discourse* that his readers not hesitate to present to him all possible objections: "Cum autem ibi [in the *Discourse on Method*] rogassem omnes quibus aliquid in meis scripturis reprehensione dignum occurreret, ut ejus me monere dignarentur, nulla in ea quae de his quaestionibus attigeram notatu digna objecta sunt, praeter duo, ad quae hic paucis, priusquam earumdem accuratiorem explicationem aggrediar, respondeo." "In the *Discourse* I asked anyone who found anything worth criticizing in what I had written to be kind enough to point it out to me. In the case of my remarks concerning God and the soul, only two objections worth mentioning were put to me, which I shall now briefly answer before embarking on a more precise elucidation of these topics."[1]

This declaration calls for two comments. First, there are two objections that precede the *Meditations:* one based on the claim that when the mind turns to itself it perceives only the *res cogitans,* the other based on the claim that the existence of a thing more perfect than myself can be deduced from my idea of it. Further, each of these objections receives a brief reply as early as this Preface. The first: "Cui objectioni respondeo," "To this objection I reply" that I do not know a thinking thing except according to the order of perception *(ordo ad meam perceptionem),* without claiming to achieve the order that follows the truth of the thing *(ordo ad ipsam rei veritatem).* The second: "Sed respondeo," "But I reply" that the idea must be taken objectively and not materially. In a word, the *Meditations* are not followed but positively preceded by objections and replies. Second, these objections (and replies) follow from a request made as early as the *Discourse (ibi rogassem);* in other words, the objections that precede the *Meditations* also proceed directly from the *Discourse,* as the conclusion of the *Discourse* confirms:

> I cannot tell if I have succeeded in this, and I do not wish to anticipate any one's judgements about my writings by speaking about them myself. But I shall be very glad if they are examined. In order to provide more opportunity for this, I beg all who have any objections to take the trouble to send them to my publisher, and when he informs me about them I shall attempt to append my reply at the same time, so that readers can see both sides together, and decide the truth all the more easily. I do not promise to make very long replies, but only to acknowledge my errors very frankly if I recognize them; and where I cannot see them I shall simply say what I consider is required for defending what I have written, without introducing any new

1. AT 7:7, ll. 14–19; CSM 2:7.

material, so as to avoid getting endlessly caught up in one topic after another.[2]

This solemn declaration amounts further, in Descartes' mind, to a commitment for him and for his eventual objectors to give reasons for their positions and nothing but reasons—in short, to do nothing but argue rationally. He also refers to this in a letter to Mersenne, in which he is in fact addressing the rector of Clermont College, Father J. Hayneuve, on the subject of the attacks of Father Pierre Bourdin: ". . . tam expresse enim in *Dissertatione de Methodo* rogavi omnes, ut me errorem, quos in meis scriptis invenirent, monere dignarentur, tamque paratum ad illos emendandos me esse testatus sum, ut non crediderim quemquam fore, qui alios erroris condemnare, quam mihimet ipsi errores ostendere, de cujus saltem charitate erga proximum non mihi liceat dubitare." "For I have indeed, in the *Discourse on Method,* asked all those who should find errors in my writings, to be so kind as to warn me of them; thus I have given assurance that I was ready to correct them, and I did not believe there would be any one who prefers to condemn me before others in my absence rather than to show my errors to me in person, especially any one professing the religious life, whose charity toward his neighbor one would not venture to call into doubt." [3]

Remarkably, it is precisely the same passage of this letter that Descartes cites when, at the close of the volume of *Meditations,* in a letter to Father Jacques Dinet, provincial of France, he complains again about the intrigues of the same Father Bourdin: "Quorum ubi fui admonitus, dedi statim litteras ad R. P. Rectorem ejus Collegii [J. Hayneuve], quibus rogabam 'ut quandoquidem opiniones meae dignae visae fuerunt quae ibi

---

2. AT 6:75, l. 19–76, l. 5; CSM 1:149–50.

3. To Mersenne, Aug. 30, 1640, in AT 3:169, ll. 2–30; unless otherwise noted, translations in this essay are by Marjorie Grene. This Latin letter to Mersenne was apparently intended to be read by the Jesuits of Clermont College. Moreover, in it Descartes takes up again his argument in an earlier letter, directly addressed to Father J. Hayneuve, doubtless dated July 22, 1640: "Praeterea profiteor me ab omni pertinacia quam maxime esse alienum, nec minus paratum ad discendum quam ullus alius possit esse ad docendum; quod jam ante etiam professus sum in *Dissertatione de Methodo,* quae meorum *Speciminum* praefatio est; ibique expresse (p. 75) rogavi omnes qui aliquid contra ea quae proponebam dicendum haberent, ne suas ad me objectiones mittere gravarentur." "Further, I profess, for myself, to be absolutely free of any obstinacy, and to be no less disposed to learn than another is to give lessons; that is what I already professed in the *Discourse on Method,* which serves as a preface to my *Essays;* for I there expressly asked all those who had anything to say against what I proposed not to refrain from sending me their objections" (AT 3:99, ll. 9–16).

publice refutarentur, me quoque non indignum judicaret, ad quem refu-
tationes istes mitteret, quique inter vestros discipulos censeri possem.
[. . .] quodque tam expresse in *Dissertatione de Methodo* pag. 75 rogarim
omnes, ut me errorum, quos in meis scriptis invenirent, monere dignarentur." "Being warned of this, I at once wrote to the Reverend Father of
the College, and begged that 'since my opinions had been judged worthy
of public refutation, he would not also judge me unworthy—I who might
still be counted among his disciples—to see the arguments which had
been used to refute them.' . . . Further, in the *Discourse on Method,*
p. 75, I asked all those who may read my writings to take the trouble of
making me acquainted with any errors into which they may have seen me
slide."[4] So, by a citation of a citation, his declaration in the *Discourse* in
which he calls for objections in order to be able to reply to them rationally, Descartes closes the whole of the *Meditations* through the letter to
Dinet. He had already used this same citation, as we have seen, in the
opening of the *Meditations* in the address to the reader. It must be concluded, therefore, that the complete text of 1641 is framed by the call for
possible objections and the promise of responding arguments that Descartes had already made at the close of the *Discourse on Method.*

This textual fact calls for several remarks. First, Descartes does not
envisage publishing his own theses without the prospect of a contradictory debate that confirms them (or possibly refutes them). In his mind
publication necessarily gives rise to public discussion, and that is doubtless why he preferred to remain semi-anonymous in 1637. But in 1641,
in the *Meditations,* he radicalized that demand by appending the record
of the debate to the first publication of the theses, something he had not
done in the *Discourse;* thus it is through the inclusion of the *Objections*
and *Replies* in the *Meditations* that Descartes consciously accomplished
what in 1637 still remained a mere intention. Consequently, and contrary
to a widespread legend, Descartes is neither here nor elsewhere anything
like a solitary, or even autistic, thinker, soliloquizing, in the manner perhaps of Spinoza. On the contrary, in every one of his works he offers to
convince his readers that he has demonstrated results; for demonstration
here constantly retains the double function of evidence: to make something manifest and to give a proof (evidence) of it to another. Cartesian
reason is communicative, precisely because truth manifests itself by a
display of evidence; indissolubly, at one and the same time, it is to

---

4. Letter to Father Dinet, in AT 7:567, ll. 20–26, and 568, ll. 4–7; trans. Haldane and
Ross, in Descartes [1911] 1967, 2:150. Cottingham omits this pasage, as did Alquié (Descartes 1963–73) and the Beyssades (Descartes 1979).

one's own reason and to the community of those looking on that the thing appears.

In short, it appears clearly that Descartes proceeded in this way for each work, following a schema that can be summarized as consisting of three installments: the plot develops in a text, then in objections, and finally in replies. For convenience, I shall call this plot the responsorial schema. Note that this schema is not limited to the writings of 1637 and 1641; in part it already governed the *Rules,* which, even though they remained unpublished, constituted throughout objections to Aristotelianism. Descartes can also be seen, paradoxically, to bring this schema to its fulfillment in the *Principles,* in which the style of the objectors, their vocabulary, and their concepts are integrated in the very exposition of the text, so that at last argument, objections and replies compose a single whole. Doubtless it is the same with the *Passions of the Soul,* where the objections and replies precede the text in the form of three exchanges of letters, so that the text acquires the appearance of a global reply. There remain the two cases of 1637 and 1641. It is in 1641 that the schema is expressed in the clearest fashion, in three plainly distinct and chronologically ordered installments; nevertheless, it is announced as such for the first time in 1637, even though at that point the objections and replies are not yet integrated with the text and remain pieces added in the correspondence.

From here on, the stages in my argument follow quite naturally. It will be appropriate to inquire first whether and to what extent the *Discourse* received objections—and what objections—and then whether and to what extent Descartes gave them correct replies. Only then can one envisage a last question: What is the link between the responsorial schema of the *Discourse* and that of the *Meditations?* Is it only a question of a repetition (even though radicalized) or of a closer link with the responsorial schema of the *Discourse?*

Objections and Replies for the *Discourse on Method*

No one denies that the *Essays* of 1637 received numerous objections. Descartes himself made a provisional catalogue of them in 1638: "As to foreigners, Fromondus of Louvain made several fairly extensive objections; and another called Plempius, who is a professor of medicine, has sent me some concerning the movement of the heart, which, I believe, contain all that could have been objected to me on that matter. Further, another, also from Louvain, who did not want to give his name, but who, between us, is a Jesuit [Ciermans] sent me some concerning the colors of the rainbow. Finally someone else from The Hague sent me some touch-

ing on various matters; that is all I have received up to the present."[5] To
the objections of "foreigners" must be added those that came from
France, principally those of Pierre de Fermat, Gilles Personne de Rober-
val, and Jean de Beaugrand dealing with the *Geometry* as well as those of
the Jesuit Bourdin concerning such material as the *Dioptrics*.[6] Descartes
replied to all these objections, not allowing for an instant any suspicion
of the slightest remorse or the slightest hesitation. From his point of view
at least, it was entirely clear that he had vanquished his adversary in each
debate. Thus we have here the simplest schema of discussion: theses, objections, and replies.

But the situation is different for the objections that concern the metaphysics of the *Discourse*. It is more difficult to examine them, since we
no longer have at our disposal the text of those sent by Jean de Silhon,
Guilliaume Gibieuf, Antoine Vatier, and others. But by a rare chance we
do have access to those that Pierre Petit presented to Descartes.[7] They are
of special interest, since, bearing on the metaphysics (the fourth part of
the *Discourse*), they evoke an ambiguous reaction from Descartes. They
are ambiguous, first, because, despite two clear and plain condemnations
transmitted to Mersenne, Descartes never formally replied to Petit (who,
moreover, complained of this to Mersenne). They are ambiguous also because Descartes hesitates in this case between two kinds of defense. First,
he claims to evade Petit's objections by saying that they consist only in
"some commonplaces borrowed for the most part from atheists, which
he piles up without judgment, pausing chiefly at what I have written
about God and the soul, in which he has not understood [the meaning
of] the words."[8] Here the adversary would seem to be opposing the theses
of Descartes (badly understood, moreover), by maintaining the point of
view of atheists; thus his harmfulness is due to his atheism, borrowed if
not explicitly his own, in the face of Cartesian metaphysics, which would
defend the cause of God. This critique of Descartes', one can infer, is
directed to two objections of Petit. In the first, Petit questions whether we
have at our disposal an idea of God. He appeals to the hypothesis that
"[a good person may be] father of an atheist or vice versa" and thus rejects the innateness of the (nontransmissable) idea of God. He also invokes the "stories that clerics themselves have told of the Canadians"
and questions whether all the "thoughts of perfections" can and must be

5. To Mersenne, June 29, 1638, in AT 2:192, ll. 9–19.
6. On these objections and replies, see, for example, Armogathe 1987; Costabel 1982;
Dibon 1990; and Rodis-Lewis 1988.
7. See Waard 1925.
8. To Mersenne, May 27, 1638, in AT 2:144, ll. 16–21.

concentrated on "a single subject" and could not be dispersed among several "gods." Moreover, he attributes the presence in us of such an idea of God to the education that "makes us join our two hands together, . . . puts us on our knees, . . . makes us beat our breasts, in a word . . . imprints us and . . . stamps us so before a divinity in the mind, that it would be to strip us of own nature or abandon our own characters to lose the idea of it."[9] These arguments can indeed seem to Descartes to be "borrowed . . . from atheists," since they anticipate the objections against the innateness of the idea of the infinite, to which he will reply in the Third Meditation.

Some months later Descartes' tactics have changed completely: Petit is no longer said to be developing reasoning borrowed from atheists, but instead Descartes sees him as using "one [argument] that he has borrowed from me." The borrowing now stems, not from adversaries, that is, from the atheists, but from Descartes himself. But then, how can he explain that Petit is still an adversary? Even though he never answered Petit, Descartes did explain to Mersenne that "the reasons he gives to prove the existence of God are so frivolous that he seems to have wanted to make game of God in writing them; and even though there is one that he has borrowed from me, he has nevertheless deprived it of all its force by putting it the way he has put it."[10] This time, Petit's fault consists in an inaccurate and caricaturelike pastiche of the correct "arguments" of Cartesian metaphysics concerning the existence of God. We can reconstruct the double line of reasoning that Descartes thus invokes. As for the argument "borrowed from me," it is probably a question of the argument in which Descartes "compares the certainty of the existence of God with geometrical demonstrations," that is, the so-called ontological argument of the Fifth Meditation; and Petit, according to Descartes, "has nevertheless deprived it of all its force" by declaring straight off that "we must not say that *the existence of God is at least as certain as a demonstration of Geometry,* since it is incomparably more certain. *Ego sum qui sum:* there is no one but himself who properly exists."[11] But, Descartes adds, Petit understands poorly what he has borrowed. We may be astonished at such a reproach, since it is striking that Petit is here maintaining to the letter a Cartesian position. He cites—without knowing it—a formula of 1630: "to demonstrate metaphysical truths in a manner that is more evident than the demonstrations of geometry." Perhaps Petit is somehow putting what was in fact the Cartesian thesis of 1630 in opposition to the

---

9. Petit, in Waard 1925, 70–71.
10. To Mersenne, Oct. 11, 1638, in AT 2:391, ll. 20–24.
11. Petit, in Waard 1925, 72.

thesis of 1637, which Descartes in the *Discourse* expresses more modestly and ambiguously: "[I]t is at least as certain that God, who is that perfect Being, is or exists, as any demonstration of Geometry could be." [12] As to the "frivolous" arguments, it is doubtless a question of the physical and cosomological proofs advanced by Petit—such as those from the "illumination of the whole Earth," the qualities of the sea, and the finitude of the elements of the world. [13]

Thus it is possible in this way to treat Petit's objections on the one hand as strictly adversarial positions or, on the other, as deformations of Cartesian doctrine—in short, now as borrowings from atheists, now as borrowings from me. Does this ambivalence rest solely and in the first instance on a lack of understanding on the part of Petit? This becomes doubtful to begin with if we consider that the Cartesian texts can be related to the two series of objections as if the definitive (printed) text had reproduced them in order to reply to them better thereafter. It becomes doubtful also when the suspicion is raised that the metaphysics of the *Discourse* itself is liable to contradictory interpretations. I have shown elsewhere that the metaphysics of 1637 was still inadequate, particularly with regard to the existence of God, since it fails to employ the concept of causality. [14] We may also suspect that Petit's objections play a part in exposing the faults of the Cartesian demonstration, which they would help to indicate. These three hypotheses have still to be discussed. Only a serious conceptual study could properly establish the metaphysical importance of the objections put forward by Petit.

Nevertheless, in all the cases and without waiting for other confirmations, it is patent that Descartes recognized several times that he could not reply definitively and immediately to objections made to the metaphysics of the *Discourse on Method*. In March 1637 he confesses to Mersenne that "as to your second objection, that is, that I have not sufficiently explained how I know that the soul is a substance distinct from the body, the nature of which is only to think, which is the only thing that makes obscure the demonstration of the existence of God, I admit that what you write about this is very true, and also that this makes my dem-

---

12. Compare letter to Mersenne, April 15, 1630, in AT 1:144, ll. 15–17, and AT 6:36, ll. 29–31. We can find here confirmation of the difference in standing between the proof from the infinite (where God's existence is demonstrated more certainly than the truths of mathematics) and the so-called ontological proof (where God sees his existence demonstrated neither more nor less certainly than the certainty with which the properties of a triangle result from its essential definition). See Marion 1986a; 1986b, chap. 5.

13. Petit, in Waard 1925, 75–79.

14. See Marion 1991.

onstration of the existence of God difficult to understand." In the same month he admits the same weakness to Silhon: "I confess that it is a great deficiency in the writing that you have seen, just as you observe, and that I have not sufficiently elaborated the arguments by which I claim to prove that there is nothing in the world which is more evident or more certain than the existence of God and of the human soul, so as to make them easy for the whole world." Finally, in February 1638 he admits to Father Vatier that "it is true that I have been too obscure in what I have written about the existence of God in this Treatise on Method, and although it is the most important section, I admit that it is the least fully elaborated in the whole work." [15] Thus the objections (even if they are misinterpretations) of Pierre Petit do not lack a certain justice: they arose, doubtless in error but not without reason, from an inadequacy in the metaphysics expounded in the fourth part of the *Discourse on Method*. For want of the argument of hyperbolic doubt, the abstraction of mind from matter is there not adequate for the idea of God (the idea of the infinite) to be perceived clearly and distinctly. Further, this is surely why, in the period following 1637, Descartes never stops underlining the particular difficulty of metaphysics, "which is a science that hardly any one understands," since "the part of the mind that helps most in mathematics, to wit, imagination, does more harm than good for metaphysical speculations." [16]

From this multiple admission, one conclusion follows clearly enough: while Descartes replied in the text of the *Essays* to the objections that were addressed to it, the text of the *Discourse,* and particularly of part 4, on metaphysics, received objections to which, according to the admission repeated by Descartes himself, it was impossible to reply. Thus the responsorial schema was achieved completely in the scientific essays but remained incomplete, in anticipation of replies, in the metaphysical essay. In other words, in 1637 the request for rational and well-argued discussion permitted the validation of the scientific results but not of the metaphysical conclusions. The question then becomes: Did Descartes ever reply to the objections left without reply in 1637?

The Essentially Responsorial Function of the *Meditations*

I propose the following hypothesis: as a whole, the *Meditations* constitute, with several years' delay, replies to the objections made to the "some-

15. To Mersenne, March 1637, in AT 1:349, l. 29–350, l. 5; to Silhon, March 1637, in AT 1:352, ll. 2–3; to Vatier, Feb. 22, 1638, in AT 1:560, ll. 7–11.

16. To Mersenne, Aug. 27, 1639, in AT 2:570, ll. 18–20; and Nov. 13, 1639, in AT 2:622, ll. 13–16. See also the letter to Mersenne of Oct. 16, 1639, in AT 2:496, ll. 22–23.

thing metaphysical" inserted in the *Discourse on Method*.[17] In other words, we should not simply speak of objections made after the *Meditations*, in the *Objections* and *Replies*, but consider the *Meditations* themselves as first and essentially replies given in 1641 to the objections formulated in 1637. Moreover, this hypothesis would confirm the view that the *Meditations* constitute a repetition, with more powerful means and more radical concepts, of the fourth part of the *Discourse*, which remains but a mere sketch.

A first argument for this hypothesis is based on a letter to Mersenne of November 1639. Declaring mathematicians incapable of thinking by means of pure understanding and of liberating themselves from the imagination, which "does more harm than good in metaphysical speculations," Descartes announces: "I have now in my hands a Discourse, in which I try to clarify what I have previously written on this subject; it will be no more than five or six printed sheets; but I hope that it will contain a good part of metaphysics." [18] What is this "Discourse"? Its function is to clarify what has been written "previously" on "this subject," that is, on metaphysics: this appears to be the metaphysics of the *Discourse on Method*, which Descartes has admitted remained "difficult to understand," because it was too "obscure." [19] This cannot be a question of the "beginning of metaphysics" of 1629, since what is necessary is precisely to reply to the difficulties of the publication of 1637; besides, the fragment of 1629, of which in any case we know nothing, could not have been reemployed except at the price of radical modifications. The text in question here is thus very probably the first sketch of the *Meditations* of 1641, understood as a repetition and a deepening of the abstraction from sense, which permits making absolutely convincing the two demonstrations sought for: that of the spirituality of the soul and that of the existence of God. It is the failure of the *Discourse* to meet the objections made to its metaphysics that, as a reply to these objections, evokes the *Meditations*.

There is more: from the start of the project of writing the *Meditations*, Descartes foresees submitting them for objections; that is to say, he presents them for a second time in a responsorial schema. In order this time to achieve full certitude and to convince completely, the replies made in 1641 to the objections of 1637 (first responsorial schema) had to be exposed to objections (second responsorial schema). Indeed, the November 1639 letter to Mersenne continues: "And in order to do this better [that

17. To Mersenne, March 1637, in AT 1:349, l. 26.
18. To Mersenne, Nov. 13, 1639, in AT 2:622, ll. 13–20.
19. To Mersenne, March 1637, in AT 1:350, l. 5; to Vatier, Feb. 22, 1638, in AT 1:560, l. 7.

is, to clarify], my plan is to have printed only twenty or thirty copies, in order to send them to the twenty or thirty more learned theologians with whom I might be acquainted, in order to have their judgment, and to learn from them what it will be well to change, correct, or add in it, before presenting it to the public." [20] Thus the connection of the *Meditations* with their *Objections* and *Replies* simply doubles, by inverting, the earlier connection of the *Meditations,* as replies, to objections made to the metaphysics of the *Discourse* and until then left pending.

A second argument in favor of this thesis comes from a letter to Constantijn Huygens in July 1640. Huygens had happened to hear about the new metaphysical work projected by Descartes. Descartes replies to him with a denial—"I have not yet placed anything in the hands of my publisher, nor have I even prepared anything that is not so slight as not to be worth speaking of"—and then admits the existence of this "something of metaphysics." And it is here that he alludes again to the two determinations we have already encountered. It is a question of a reply to the objections made to the *Discourse* and of a request for examination by other objectors: "[I]n short, nothing can have been reported to you that is true, if it is not what I remember saying to you last winter, namely, that I intended to clarify what I wrote in the fourth part of the *Method,* and not to publish it, but to have only twelve or fifteen copies of it printed, in order to send them to twelve or fifteen of the chief of our theologians and to await their judgment." [21] Here again, more than six months after a similar declaration to Mersenne, Descartes promises first to "clarify" what the fourth part of the *Discourse* contains that is "too obscure," which amounts to furnishing at last the long-delayed reply to the objections of 1637. Here again Descartes announces that he will submit his reply (of 1641) to the objections (of 1637) to a new round of objections, those of "twelve or fifteen . . . theologians" (no longer twenty or thirty). Thus the responsorial scheme is doubled, since the *Meditations* here play the part, first, of replies to the objections made to the *Discourse,* in order then to take on the role of the text to which objections are to be made. But there is more: the copies of the new treatise on metaphysics sent for examination to the eventual objectors will not be sold or circulated to the public: "I intend . . . not to publish it, but to have only twelve or fifteen copies . . . printed." It must be understood that the exposition of the reply to the objections of 1637, that is, the *Meditations* themselves, hold from the

---

20. To Mersenne, Nov. 13, 1639, in AT 2:622, ll. 20–26.

21. To Huygens, July 1640, in AT 3:102, ll. 5–16. Adam and Tannery state that we have no trace of what was "said last winter."

start a relation to the future objections so essential that their printing remains limited to a restricted circulation, exactly as if they were not supposed to be accessible, nor could be understood, without the dialogue to come through objections and replies. The printing limited to restricted circulation would even allow us to consider this a simple printing *pro manuscripto,* neither definitive nor public. The publication, strictly speaking, would not take place until the objections had been received and refuted by proper replies. In short, to do what he had left pending in the *Discourse*—replying to the objections on the metaphysics—Descartes in the *Meditations* goes so far as to include the whole responsorial apparatus in the original text.

A third argument is furnished by a letter to Henricus Regius of May 24, 1640. It gives evidence of a reading of the *Meditations* and of objections made to them even before their appearance (and certainly before publication).[22] Descartes confirms the existence of this reading by thanking his correspondents for it: "Multum me vobis devixistis, tu et Cl. D. Aemilius, scriptum quod ad vos miseram examinando et emendando." "I am much obliged to you, you and M. Emile, for examining and correcting the writing that I submitted to you." Not only did they deign to "interpunctiones et orthographiae vitia corrigere," to "correct the punctuation and the errors of orthography," but they also presented "objections." It is important to study the first of these objections: we have at our disposal something of the wisdom, the power, the goodness, the magnitude, etc., starting from which we can form the idea of a wisdom that is infinite or at least indefinite, etc. Descartes replies, obviously, that this formation (of the idea of infinite wisdom and so on) by the passage to the infinite of qualities that are finite in us would remain impossible if we did not have at our disposal, from the start and innately, the idea of the infinite and of the infinite in act. But it is remarkable that what is in question is, almost in the same terms, the objection put forth in 1637 by Pierre Petit and soon again to be repeated by Gassendi.[23] We can even conjecture that a development of the Third Meditation was directly

22. "In April 1640 Descartes had completed the composition of the *Meditations*" (Alquié, in Descartes 1963–73, 2:171). He had completed the composition, certainly, but not yet the definitive text, which would integrate many remarks and improvements suggested by readers both favorable and critical.

23. To Regius, May 24, 1640, in AT 3:63, l. 2–64, l. 20. Baillet gives the following version of this exchange: "[H]e had allowed his manuscript to be read by several friends in Utrecht who had earnestly requested it, and particularly by Messieurs Regius and Emilius, who had been charmed to ecstasy by it. . . . they proposed to him two difficulties concerning the idea we have of the infinite and infinitely perfect Being, and asked him for a fuller elucidation of

inspired by Descartes' concern with replying to the objections of Petit in 1637 and Regius and Emilius in 1641.[24]

Let us then consider it as established that the composition of the *Meditations* was governed by two types of objections: those addressed to the fourth part of the *Discourse on Method* but, by Descartes' own admission, left without replies, and then in turn those of the readers of the manuscript (or of a limited edition). It remains to identify in the two cases what modifications of the initial text the objections provoked. This enterprise is easier for the readings of the manuscript of the *Meditations* before its definitive publication. We need only think of the numerous corrections of the text printed in 1641 that were provided by Descartes at the suggestion of Mersenne or in discussion with him.[25] As for the replies to the objections provoked by the *Discourse,* they cannot be found at all in the text of the *Meditations* except by a strict comparative examination of the conceptual situations of the two works. Note first that this investigation would give to the differences between the metaphysics of 1637 and of 1641—indisputable in my view—a new status and a new interest: it would no longer be a question of fearing contradictions and incoherences between two equally Cartesian texts, but rather of reconstituting a deepening of Cartesian thought from one to the other, through integration, in the second exposition of metaphysics, of replies to the objections made the first exposition and left without reply. On this hypothesis, the strict corpus of the six meditations ought to be read, indissolubly, as an ensemble of replies to the scattered objections made to the *Discourse on Method* and as a text itself destined from the first—even before its (regular) publication—to be submitted to objections, to which Descartes

---

what he had written in his Treatise. M. Descartes was pleased to accord them this satisfaction" (1691, 2:103).

Three points need correction: first, the manuscript was made public before printing, and thus corrections suggested by the objections could still be assimilated; second, it was Descartes who sent (*miseram,* AT 3:63, l. 3) the manuscript that the correspondents had "earnestly requested" of him; and third, the second objection bears less on God than on the status of evidence as long as God remains unknown.

24. See AT 7:46, l. 29–48, l. 2.

25. After the manuscript was sent to Mersenne, however—that is to say, during the printing—Descartes still modified his text. Sometimes, indeed, his changes concerned only style (to Mersenne, Dec. 24, 1640, in AT 3:267, l. 22–268, l. 8). More frequently, however, they affected the basis of the argumentation (to Mersenne, Dec. 31, 1640, in AT 3:271, l. 9–274, l. 24, and March 18, 1641, in AT 3:334, l. 10–335, l. 10). Sometimes he corrected the *Replies* themselves (to Mersenne, March 4, 1641, in AT 3:329, l. 7–331, l. 9, and March 18, 1641, in AT 3:335, l. 11–338, l. 5).

would reply.[26] Not only would it be illegitimate to read the *Meditations* in abstraction from the *Objections* and *Replies,* with which they intentionally form an organic whole, but it would also be wholly illegitimate to read them otherwise than as replies to the objections evoked by the *Discourse on Method.* Far from being soliloquy or solipsism, Cartesian thought, insofar as it obeys a logic of argumentation, is inscribed in its very origin in the responsorial space of dialogue.

26. As soon as Descartes finished with the composition of the *Meditations* and *Replies,* faithful to his argumentative technique (theses, objections, replies), he formed the project from which the *Principles of Philosophy* would eventuate—"comparison of these two Philosophies" (to Mersenne, Nov. 11, 1640, in AT 3:233, ll. 14–15), "in such a way that we will easily see the comparison of one with the other, and that those who have not yet learned the philosophy of the School will learn it much more easily from this book than from their masters, because they will learn by the same means to distrust it" (to Mersenne, Dec. 1640, in AT 3:259, l. 27–260, l. 4). It is only in a letter to Father Charlet in December 1640 that Descartes proves somewhat misleading in attributing to "one of my friends" the project of "a full comparison of the philosophy that is taught in our schools with the [philosophy] I have published, in order that by showing what he thinks bad in the one, he will make it so much the clearer what he judges to be better in the other" (AT 3:270, ll. 4–9).

THEO VERBEEK

————————— ᘒ 2 ᘗ —————————

THE FIRST OBJECTIONS

Of all the objections to Descartes' *Meditations,* none perhaps have been so neglected as the First.[1] Written by the Dutch archpriest Johannes Caterus (1590–1655) and limited to a discussion of some specific points of natural theology, they invite less interest than, for example, the Fourth and the Fifth Objections, which not only cover a great variety of subjects but were also written by celebrities such as Arnauld and Gassendi. Not surprisingly, therefore, information about their author is scant, virtually limited to what is found in Adrien Baillet's *Vie de Monsieur Descartes.*[2] In this essay I will present the few additional data I have collected, discuss the First Objections, and conclude with some remarks on the parallels between Caterus' objections and some Orthodox Calvinist critiques.

About Caterus

All that was claimed by the Union of Utrecht (1579), generally regarded as the first manifestation of Dutch political identity, was that each of the seven provinces would be free to rule in matters of religion.[3] At any rate, since the aim of the Dutch revolt was religious freedom rather than the replacement of one creed by another, there was no legal ground to persecute the Catholics.[4]

In 1559, however, a concordat had been concluded between Spain and the Holy See, by virtue of which the Spanish king was to appoint bishops in Utrecht, Haarlem, Middelburg, Deventer, Groningen, and Leeuwar-

1. The only studies are Monchamp 1886, 124–34; and Caraballese 1946, vol. 1.
2. Baillet 1691, 2:110; cf. Monchamp 1886, 124.
3. Cf. Knuttel 1892–94, 1:1. The point was to play an important role during the Remonstrant Crisis when the "grand pensionary" of Holland, Oldebarneveld, challenged the right of the States General to rule in church matters.
4. Cf. Knuttel 1892–94, 1:2.

den.[5] After the abjuration in 1572, bishops and priests were thus seen as agents of the enemy. Anti-Catholic feelings were also aroused by the Rennenberg plot (1580) and the murder of William of Orange (1584), the latter especially because the murderer, Balthasar Gerards, admitted that he had acted on Jesuit instigation.[6] Before long Catholics became the object of repressive measures. Religious orders and corporations were forbidden. Priests were threatened with exile. People attending religious meetings were fined. The use and display of ecclesiastical garments were prohibited, as were all meetings with a religious character.[7]

Humiliating though these measures were, the very fact that they had to be renewed frequently shows that they were ineffectual. Most of the time the authorities intervened not because Catholic religion was considered false but because they wanted to prevent scandal and social trouble. Religious services were interrupted only if the number of attendants was particularly great or if the *klopjes*—literally "knocksters," pious women who went around knocking on the doors of Catholics to announce that mass was going to be read—acted too visibly. Priests were left in peace as long as they did not try to convert Calvinists or incite Catholics to plot against the civil authorities. In general, mass could be said if people wished it, provided that it was held in a private house.

At any rate, a large part of the population, especially in the countryside, remained Catholic.[8] In the Noorderkwartier, the upper part of the Province of Holland with the town of Alkmaar as its center, for example, the population was basically Catholic. Almost any country village had its priest and its semi-clandestine church. Egmond, for example, Descartes' favorite dwelling place, was never without a priest. In Alkmaar, which at the time had probably no more than seven thousand inhabitants, there were five thousand communicants, attended to by five priests, two of whom were secular, one Jesuit, one Minorite, and one Dominican.[9] A price had to be paid for this activity, though. Catholics had no legal status and could be surprised at any time, and they also had to pay substantial bribes. In 1639 the Utrecht police entered the house of Hendrika van

5. Cf. Rogier 1947, 1:201–26.

6. Georges (de) Lalaing, called Rennenberg, was a Catholic friend of Orange, who in 1580 bargained away the northern provinces. Cf. Trosee 1894.

7. Knuttel 1892–94, 1:2–4.

8. This point is controversial. Catholic historians, like the late Rogier, held that the country remained basically Catholic and that the official politics of "Protestantization" failed, but their conclusions were contested by others. It seems safe to admit that the number of Catholics was greater than it is generally assumed to have been, probably 50–60 percent.

9. Torre 1882–84, esp. 12:422.

Duvenvoirde in an attempt to arrest Philippus Rovenius, the apostolic vicar. Although Rovenius escaped to Amsterdam, his secretary, Godfried van Moock, was arrested, and his library was confiscated.[10] A confirmation ceremony in Zijdewind, a small village of the Noorderkwartier, attended by several thousand Catholics—a substantial number in an area that was sparsely populated and that, before the reclamations, was dotted with lakes—was disturbed first by the local populace and then by the police. The church was demolished, and the goods of Jacobus de la Torre, Rovenius' coadjutor, were confiscated.[11] Cases like these were rare, however. If Catholics proceeded with discretion, they were left in peace. Within a year of Rovenius' flight to Amsterdam, he returned to Utrecht. Although the church of Zijdewind was destroyed, none of the local priests, each of whom was known by name to the police, were arrested. The fact that some people earned a substantial living by accepting bribes from the Catholics was a motive not to act so severely that the Catholics would go elsewhere. Otherwise, few people really cared.

Born in Antwerp in 1590, Caterus (also called Cater, de Cater, de Caters, or de Kater) probably belonged to a Dutch family who had fled to the Spanish part of the Low Countries.[12] A younger brother, Jacob, born on December 9, 1593, entered the order of the Jesuits in 1611 and died rector of the College of Courtray in 1657. His sermons, for which he seems to have been particularly famous, were collected and published.[13]

Caterus matriculated in Louvain, on February 28, 1620.[14] It is not known whether he lived in the newly founded seminary of the Haarlem clergy, the Pulchery, but given his later connection with the Haarlem diocese, it seems likely.[15] If so, he also met Cornelius Jansenius, president of

10. The books were deposited in the library of the newly founded university. On Rovenius, see Visser 1966; Rogier 1947, 2:71–73.

11. Poelhekke 1964.

12. There is some uncertainty in the literature: Baillet says that he was born in Antwerp (1691, 2:11–12), and Heusden says that his family came from Alkmaar but that he was born in Antwerp (1714, 2:418–21). According to the matriculation registers in Leuven, however, he was "Harlemensis." Since his younger brother, Jacob, was certainly born in Antwerp, it is most likely that, having been forced to move to Antwerp, the family continued to see themselves as inhabitants of their native town, Haarlem.

13. Sommervogel et al. 1890–1932, 2:876–78 (lists ten titles that may possibly be attributed to Jacobus de Cater); cf. *Bibliotheca Catholica Neerlandica Impressa* 1954, no. 9850 (*Virtutes cardinales ethico emblemate expressae* [Antwerp: Moretus, 1645]).

14. *Matricule de l'Université de Louvain* 1962, 5:58 (no. 289). He is classified with the "rich students," that is, those who studied without a scholarship, and as a member of the College of the Swine.

15. The college was called after a statue of the Holy Virgin placed above the entrance, which the people of Louvain called "the beautiful Virgin." Cf. Vregt 1880.

the college from 1616 until 1624 and author of *Augustinus,* although Jansenius' correspondence shows no trace of a meeting. On January 4, 1629, Caterus reported himself to the Amsterdam magistrate, saying that, having been abroad, he was unable to do so earlier, which suggests that he remained in Louvain until the end of 1628.[16] On the Amsterdam list Caterus is designated as "w.p." (*wereldlijk priester,* that is, secular priest) and "Mr" (*Meester,* that is Master of Arts). He matriculated again at Louvain in 1634.[17] The reason for this was probably his appointment to the Haarlem Chapter in 1632; the rules of the Chapter specified that of the ten prebends, one belonged to the bishop and was part of the *mensa episcopalis,* three belonged to doctors of theology, three to doctors of law, and the remaining three to noblemen of the diocese.[18]

In 1638 Caterus became archpriest in Alkmaar. The office of archpriest had been introduced by Rovenius, in order to preserve what might remain of pre-Reformed hierarchy. The archpriests may be compared with deans, and they governed parts of the diocese more or less independently. Little is known of Caterus' work as an archpriest. In 1641 his name appeared on the official list of priests and ecclesiastical persons that was submitted to the burgomasters of Alkmaar. In 1643 he placed a miraculous relic of the Holy Blood into a silver shrine, and in 1653 he was denounced to the burgomasters for having administered the Last Sacrament to a Reformed citizen.[19]

Caterus always fiercely defended the rights of the Haarlem Chapter, especially against the usurpation by the apostolic vicar.[20] The Haarlem Chapter was the only remnant of the pre-Reformed hierarchy of the Roman Catholic Church in Holland.[21] Since the last bishop of Haarlem, Govert van Mierlo, had had no successor, the Chapter governed the diocese *sede vacante.* The apostolic vicar, however, claimed that in 1602 the

---

16. "Mr. Jan Cater, van Haerlem, w.p., gelogeert t.h.v. Pauls Laurensz., boxenmaker op 't Water, heeft (door syn uytlandigheyt verhindert geweest sijnde) nu eerst den 4. Januarij 1629 zijnen naeme gedaen aangeven door Henrick Barentsz boekverkoper." Roever 1893; cf. Knuttel 1892–94, 1:121–23. He reported himself to the magistrate in order to obey an edict of the States General of 1622, requiring all religious persons to be officially registered.

17. *Matricule de l'Université de Louvain* 1962, 5:287 ("reintitulatus").

18. Graaf 1887, 7; cf. Mous 1962.

19. Bruinvisch 1905; Rijkenberg 1896; Bruinvisch 1905, 253.

20. Heusden 1714, 2:418–21. He did so in close concert with his friend Baldwin Cats (whose name, like that of Caterus, derives from the Dutch *kat,* or "cat"), dean of the Chapter. People said that "de kat en de kater waken over de rechten van het kapittel."

21. The Utrecht Chapters still existed but had become protestant. This is the reason why a "vicaric council" (instead of the Chapter) assisted the apostolic vicar in governing the diocese.

pope had given him the exclusive right to govern all the provinces under the jurisdiction of the States General. In his view, the Chapter was simply an anachronism. In 1616, however, the Chapter and the apostolic vicar agreed on three points: the Haarlem Chapter recognized the apostolic vicar as the supreme authority in all the dioceses of the Low Countries, the apostolic vicar recognized the Haarlem Chapter as lawful, and the Chapter was to have the right to appoint priests in the diocese, whereas the dignities of the Chapter were in the gift of the apostolic vicar.

This has all the marks of a compromise, if only because the first two articles are contradictory. Moreover, there is reason for doubt whether, under the circumstances, the third article meant very much. As long as the personal relationship between the apostolic vicar and the members of the Chapter was good, however, the agreement offered a reasonable working basis. This changed after Rovenius' death. Rovenius' coadjutor, Jacobus de la Torre, who succeeded him in 1651, refused to renew the convention of 1616, but was forced to do so because of his own financial problems. Under Zacharias de Metz, who was appointed coadjutor after de la Torre went insane, these conflicts became more serious, and they culminated in 1661, when de Metz prohibited the *praetensum capitulum* from appointing *praetensos capitulares*. A few days later de Metz officially dissolved the Chapter, an act for which he justified himself in a long memorandum to the Congregation of the Propaganda. However, before the Congregation could reach a decision, de Metz died, which allowed the Chapter to defend its rights more successfully. A definitive decision was delayed, but the Congregation continued to speak of the *praetensum capitulum,* and this way of referring to the Chapter remained in use until the restoration of the Catholic hierarchy in the Netherlands in 1853.[22]

Caterus' correspondence, kept in the archives of the Oud-Bisschoppelijke Clerezie, shows that he was a diligent priest. In fact, several times he was considered for appointment as coadjutor to the apostolic vicar.[23]

The First Objections

The First Objections to Descartes' *Meditations* take the form of a letter to two other members of the Haarlem Chapter, Bannius and Bloemaert (or Blommert), who, being personal friends of Descartes, were first ap-

22. Rogier 1947, 2:90–91.
23. The archives of the Oud-Bisschoppelijke Clerezie (or Oud-Katholieke Kerk, comprising those Catholics who separated from Rome at the beginning of the eighteenth century) are now kept in the Rijksarchief in de Provincie Utrecht (Utrecht).

proached for a reaction. The last phrases of Caterus' objections even suggest that his performance as an objector was not voluntary. At any rate, he does not seek to continue the debate with Descartes, happy as he is "to avoid a second defeat."[24]

In his objections Caterus concentrates on the theological aspects of Descartes' metaphysics, especially his proofs of the existence of God. Five points emerge from this discussion. The first is Caterus' inability to think of "ideas" in the Cartesian sense, that is, as entities of a kind that require a cause for their existence; the second relates to Descartes' conception of God as *causa sui,* or *ens a se;* the third concerns Descartes' claim that we have the idea of infinite being; the fourth involves the so-called ontological proof of God's existence, which Caterus rejects; and the fifth explores Descartes' real distinction of body and mind.

Caterus starts by summarizing the Second Meditation and the beginning of the Third. He has no specific objections to the *cogito,* but he hesitates as soon as Descartes starts to prove the existence of God by inquiring into the causes of our ideas. "What sort of cause does an idea need?"[25] In the *Meditations,* Descartes does not take the trouble to define *idea* properly. Wondering, at the beginning of the Third Meditation, what he actually knows, Descartes mentions "the ideas, or thoughts" of things.[26] A bit later he reserves the name of *idea* for images: "Some of my thoughts are as it were the images of things, and it is only in these cases that the term 'idea' is strictly appropriate."[27] Caterus, in turn, speaks of ideas as things thought of or things with objective reality: "Indeed, what is an idea? It is the thing that is thought of, insofar as it has objective being in the intellect."[28] As a result, he does not think of objective reality as the property of an idea, which should be explained accordingly, but as "merely an extraneous label which adds nothing to the thing itself."[29] After all, a thing is what it is, whether it is actually thought of or not. Accordingly, there is no reason to inquire into the cause of objective reality: "On the contrary, this requires no cause; for objective reality is a pure label, not anything actual. A cause imparts some real and actual influence; but what does not actually exist cannot take on anything, and so does not receive or require any actual causal influence. Hence, even if I have ideas, there is no cause for these ideas, let alone some cause which is greater than I am, or which is infinite."[30]

24. AT 7:101, 9:80; CSM 2:73.
25. AT 7:92, 9:74; CSM 2:66.
26. AT 7:35, 9:27; CSM 2:24.

27. AT 7:37, 9:29; CSM 2:25.
28. AT 7:92, 9:74; CSM 2:66.
29. AT 7:92, 9:74; CSM 2:67.

30. AT 7:92–93, 9:74; CSM 2:67. I have altered CSM's "though I have ideas" to "even if I have ideas," since I am not sure that Caterus commits himself to the existence of ideas.

Caterus' second problem was suggested by Descartes' second proof of God's existence. To my mind, the point of the second proof is not so much to prove that there is a God as to prove that I am not the God whose existence was demonstrated in the first proof.[31] Indeed, "if I derived my existence from myself, then I should neither doubt nor want, nor lack anything at all; for I should have given myself all the perfections of which I have any idea, and thus I should myself be God."[32] At any rate, the meaning of this argument in Descartes is not entirely clear, and Caterus is probably right when he lumps it together with the first proof. It is not so much an independent proof of the existence of God as an attempt to obviate some of the consequences that might be attached to the first proof. One of these is indeed that, given the fact that God is objectively in me and if objective reality is a kind of being that requires a cause, which is what Descartes means by saying that we have an idea of God, then it is difficult not to think that, in some way or another, we are God. This position to contemporary thinkers was exemplified in what was called enthusiasm.[33]

Caterus feels more at home when Descartes speaks plain causal language, without using it to determine the cause of an objective reality. His real point, however, relates to God's being a *causa sui* or *ens a se*. In the *Meditations* Descartes does not use these exact expressions. In order to refute the hypothesis that I myself am God, he introduces the alternative supposition that I derive my existence from myself. According to Descartes, this hypothetical possibility requires an active force, which entails that, being now in existence, I would be able to make myself also exist in the following moment. In any case, the cause of my existence can be either *a se*, that is from myself, or *ab alia*, that is, from another cause:

> [T]herefore whatever kind of cause is eventually proposed, since I am a thinking thing and have within me some idea of God, it must be admitted that what caused me is itself a thinking thing and possesses the idea of all the perfections which I attribute to God. In respect of this cause one may again inquire whether it derives its existence from itself or from another cause. If from itself, then it is clear from what has been said that it is itself God, since if it has the power of existing through its own might, then undoubtedly it also has the power of actually possessing all the perfections of which it has an idea—that is, all the perfections

31. Cf. Third Meditation, in AT 7:48–51, 9:38–40; CSM 2:33–35.

32. AT 7:94, 9:75; CSM 2:68.

33. Cf. Schoock, *Admiranda methodus* (1643) 4.2, in Descartes and Schoock 1988, 312–14.

which I conceive to be in God. If, on the other hand, it derives
its existence from another cause, then the same question may be
repeated concerning this further cause, namely whether it derives
its existence from itself or from another cause, until eventually
the ultimate cause is reached, and this will be God.[34]

Caterus does not explicitly reject anything of what Descartes says, but he
wonders in what sense the expression *a se* must be understood, positively
or negatively:

In the first, positive sense, it means "from itself as from a cause."
What derives existence from itself in this sense bestows its own
existence on itself; so if by an act of premeditated choice it were
to give itself what it desired, it would undoubtedly give itself all
things, and so would be God. But in the second sense, "from
itself" simply means "not from another"; and this as far as I remember, is the way in which everyone takes the phrase. But now,
if something derives its existence from itself in the sense of "not
from another," how can we prove that this being embraces all
things and is infinite?[35]

Descartes' reasoning is therefore inconclusive unless he is prepared to take
the expression *causa sui* in an unusual sense. Accordingly, Caterus suggests, Descartes should not "hide his meaning."[36] It must be emphasized,
however, that Caterus does not commit himself.

Caterus' third problem concerns the possibility for a finite being to
know the infinite.[37] The point is not a bad one. It was essential to Descartes that we have some positive idea of an infinite being. On the other
hand, he could not reject the principle advanced by Caterus, that the infinite qua infinite cannot be known by a finite intellect.

Caterus' next problem involves the ontological proof. This proof had
never been very popular, not because people wanted to deny that God
necessarily exists, but because they felt that it could not be used as a
proof. The reason is given by St. Thomas, whom Caterus quotes: For the
proof to be effective, it is necessary that we already have a notion of God,
if not the knowledge of his existence. As a result, the proof is not inconclusive in itself, but because of the premises that it requires, it is useless
against those for whom it is needed, that is, the atheists. All they have to
do is simply deny that they have the notion in question. Therefore, the
cosmological argument, which makes use of a shared experience, is much
more effective in convincing the atheists: "Cum Deus sit suum esse, et

34. AT 7:49–50, 9:39; CSM 2:34.   36. AT 7:95, 9:76; CSM 2:68.
35. AT 7:94–95, 9:76; CSM 2:68.   37. AT 7:96–97, 9:77–78; CSM 2:69–70.

quidnam sit nos lateat, haec propositio, Deus est, per se nota secundum se est, licet non quoad nos." St. Thomas does not deny that we have some natural notion of God, and on this point too he is quoted and followed by Caterus. God is our beatitude. And since we strive to attain beatitude, we must have some notion of God. However, this is not knowing God *simpliciter,* but knowing him *sub quadam confusione.*[38] Caterus adds an argument of his own, which amounts to stating that existence is essential to any existing being. Thus, for example, existence is an essential ingredient of the composite "existing lion." However, the fact that to an existing lion it is essential that, in fact, it exists does not mean that the lion exists at any moment. Its existence is necessary only for the time it actually exists. Again, even if we do admit that God's essence entails his existence, we still cannot know that he actually exists, unless we actually admit his existence. In brief, the fact that, if God exists, he exists necessarily does not entail that, in fact, he exists.[39]

Finally, we can deal briefly with Caterus' last problem, which relates to Descartes' real distinction of body and mind. It is clear, I think, that Caterus refuses to see body and mind as substances, probably because in good Thomist fashion he believes that the soul is the substantial form of the body. His argument, however, is taken from John Duns Scotus, who, by introducing the concept of a "formal and objective distinction," provides the conceptual means to conceive of body and mind distinctly, without having to see them as substantially different.

In sum, although Caterus' discussion is marked by some slight misunderstandings, especially with respect to the representational model of perception and so with the exact meaning of the term *idea,* his remarks are very much to the point. The argument about existence, for example, seems original to me and anticipates Kant's argument about the impossibility of conceiving of existence as a predicate. Caterus also makes some remarks critical of Francisco Suárez, who had advanced the thesis that anything that is the cause of itself is necessarily infinite.[40]

Calvinist Parallels

Caterus is the only Catholic participant in the Dutch debate on Descartes. It is interesting that many of his arguments would be used eventually by some of the other participants, who, of course, were Calvinists. This is

38. *Summa theologica* 1, q. 2, art. 1, ad. 1.
39. AT 7:97–100, 9:78–80; CSM 2:70–72.
40. AT 7:95, 9:76; CSM 2:69.

not entirely a coincidence. Thomas Aquinas had great influence in Calvinist theology during the first half of the seventeenth century and was often preferred to Suárez, who had a dangerous reputation of Pelagianism.[41] Among the problems raised by Caterus, however, one was very actively discussed, and accordingly, it represents a kind of general stumbling block for theologians, comparable to that of the necessity of ideas for philosophers. That is, of course, the idea of *causa sui* in the positive sense. It is certainly Caterus' merit to have highlighted this concept, which evidently had not concerned Descartes very much and to which he had given little attention.

The question of God as *causa sui* was taken up first by Jacobus Revius (1586–1658) and later by Adam Stuart (1591–1654). Stuart reacted to this idea in a disputation of 1647, which was commented upon by Descartes in the *Notae in programma quoddam* (1648), which in turn elicited a response by Stuart, *Notae in notas nobilissimi cujusdam viri in ipsius theses de Deo* (1648).[42] Stuart's reaction, however, is not very original and leans heavily on Revius' discussion of the point.

Nowadays Revius is mainly known as a poet, but at the time this one-time minister of the church of Deventer was probably better known as a theologian.[43] In 1641 he became the regent (dean) of the Statencollege, a college at Leiden University, founded by the States of Holland. His duty was to train students in theology, which resulted in a series of disputations covering the entire field.[44] Apart from that, he produced several works of a more specifically philosophical nature, especially *Suárez repurgatus*, an annotated syllabus of Suárez' *Metaphysical Disputations* that was published in 1643. Revius was certainly aware of Descartes. He was a friend of Henricus Reneri, who was on familiar terms with the French philosopher.[45] He probably even tried to convert Descartes to the Calvinist creed, an attempt that must have left him with few illusions.[46]

41. Cf. Ruler 1991.

42. See AT 8-B: 365–69 (I have shown elsewhere that Descartes' adversary here is not Revius but Stuart). The only surviving copy of the Stuart pamphlet is in the Bibliothèque Nationale.

43. For an English translation of some of his poetry, see Revius 1968. On Revius in general, see *Nieuw nederlandsch biografisch Wordenboek*, edited by P. C. Moljuysen and P. J. Blok, 10 vols., (Leiden, 1911–37), vol. 6, cols. 1174–76; Meyjes 1895; and Smit 1928.

44. Revius 1642–46.

45. A poem in four languages—Greek, Latin, French, and Dutch—praises Reneri's appointment at Deventer's illustrious school; a poem on the military victories of the year 1633 is dedicated to him. See Revius [1930–35] 1976, 2:112–16, 153. Revius is mentioned several times in Reneri's correspondence with De Wilhem (Reneri n.d.).

46. See Adam 1910:345. Descartes is said to have answered that he preferred the religion of his king and, when the other insisted, that of his wet nurse.

In *Suárez repurgatus* Revius had already made some remarks on Descartes, and even on the idea of God as *causa sui*.[47] However, since that work seems to have remained unsold—the publisher tried to resell it with a new title page a few years after its initial publication—Revius tried again in a series of disputations beginning in 1646.[48] Revius seems to have intended to discuss extensively every aspect of Descartes' philosophy. In fact, only five successive disputations, defended between February 4 and March 30, 1647, were devoted to Descartes' philosophy, possibly because the States of Holland, to which Revius was directly responsible, prevented him from carrying out his design.[49]

In the first disputation Revius analyzes the method of doubt. It is a great sin to doubt the existence of God, and a much greater sin to deny God. In fact, it can never be justified, not even for the time being or once in our lifetime.[50] In the second disputation Revius discusses the various means of knowing God. Scripture testifies on every page to God's existence, but natural reason too may be used. It is impossible, however, to demonstrate a priori, that is, from his cause, that there is a God, because God's being has no cause. In the third disputation Revius explicitly deals with Descartes' proofs of God's existence.[51] All Cartesian proofs rely on the fact of our having an idea of God. According to Revius, this is a form of blasphemy, since Descartes defines *idea* as "image" or "resemblance."[52] But there can be no image of God. Descartes might of course reply that we affirm or deny all kinds of things with respect to God, but according to Revius, this is no proof that we have an idea of God. Judgments, notions, or concepts are not ideas. Neither is the power to elicit ideas—an interpretation Descartes ventures on in his Letter to Voetius—an idea.[53] It is also false that fear and will are ideas, as Descartes says in the Third Meditation; that we have a "positive and real idea" of God or of an eminently perfect being; that our intellect can form the idea of an infinite thing; that we have the idea of an infinite number; or that it is possible to have a "negative idea" of "nothingness." Revius ends by quot-

47. Revius 1643, 207–9, 504–7, 518–19, 872.
48. Revius 1646–53 (the only complete copy is in the Koninklijke Bibliotheek at The Hague). For a complete list of the subjects treated, see Meyjes 1895, xxx.
49. This at least is suggested in an anonymous pamphlet by Philalethus Eleuterius Atheniensis (Atheniensis 1648). It is more or less admitted by Revius himself in *Statera philosophiae cartesianae* (1650).
50. See First Meditation, in AT 7:17, 9:13; CSM 2:12.
51. An earlier version of the same argument appears in Revius 1643, 518–19.
52. Cf. AT 7:37, 9:29; CSM 2:25.
53. AT 8-B: 166–68; Descartes and Schoock 1988, 386–87. See also *Notae in programma quoddam*, in AT 8-B: 357–58; CSM 2:303–4.

ing Suárez' definition of *idea,* according to which ideas are *causae exemplares,* that is, the mental representations an artisan uses in his work.[54] Applied literally to God, this is of course absurd. As a result, Descartes' conception of the term *idea* is confused if not simply false.

In his two other disputations Revius deals with the idea of God insofar as he is a self-caused being or *ens a se.*[55] Revius takes up the discussion in Descartes' reply to Arnauld that the expression that God is "by himself" only means that God's essence is positive. Thus, the causality of God as *ens a se* would be formal instead of efficient.[56] Revius concludes that Descartes is the first philosopher who uses the expression *ens a se* in a positive sense. He spells out the various implications of Descartes' words, asserting that their real meaning is that, with respect to himself, God is literally an efficient cause. Revius agrees with Arnauld that no theologian can endorse this conception of God.[57]

In his reply to Arnauld, Descartes admitted that it would be wrong to believe that the Son is caused by the Father because that would imply that the Son is less than the Father. If, however, the question is about God as a unitary being, there is, Descartes said, no problem at all. Indeed, if it were not permissible to ask of everything that exists why it exists, it would be impossible to prove God's existence.[58] Revius admits that some of the Church Fathers used the word *cause* somewhat loosely when speaking of the relation between the different persons of the Trinity, but none of them used that word in any sense if the question was about God's being as a unity. Of course, Revius says, efficient causality is the only means to prove that God exists, but nobody ever used the principle of causation in order to localize the efficient cause of God's being.

Descartes had also tried to avoid the problem by emphasizing that he meant to speak of the formal cause of God, that is, God's essence. Revius had no patience with that. As far as the formal cause is concerned, nothing cannot be said to be "from itself," because any being has its own formal cause and not that of another. The point, however, is that for the existence of a finite being the cooperation of an efficient cause is indispensable, whereas God does not need this. It is absurd, therefore, to affirm that God's formal cause is the efficient cause of his existence.

It is clear that Descartes underrated the problems connected with his

54. Suárez 1856–78, 1:900.
55. The text was reprinted in Revius 1648, 121–35. An earlier version of the same argument can be found in Revius 1643, 504–7.
56. Fourth Replies, in AT 7:231–45, 9-1: 179–89; CSM 2:162–71.
57. Fourth Objections, in AT 7:213, 9-1:166; CSM 2:149–50.
58. Fourth Replies, in AT 7:239–45, 9-1:183–86; CSM 2:167–70.

idea of God as *ens a se* or *causa sui,* but it is also clear that he stood by his idea that God's being has a cause of some kind. The objections of Descartes' adversaries, on the other hand, were mainly religious. According to them, it is pointless and even sinful to scrutinize the cause of God's being. Calvinist theology declares that God reveals his will in the Bible and his works through nature. To apply to God the categories of human thinking amounts to blasphemy. Philosophically, the point is somewhat different. Causality is first and foremost seen as the expression of the finitude of a being. This is clearly what Revius has in mind when he says in *Suárez repurgatus* that being as such has no cause and that to ask for the cause of infinite being is contradictory.[59]

59. Revius 1643, 12; see also 156–70.

ᔧ 3 ᔤ

CATERUS' OBJECTIONS TO GOD

In many ways, the so-called First Objections have a very special status. Their author, Johan de Kater (Caterus) is little known, and these few pages are his only known philosophical writings; they stand at the head of the *Objections,* and they were circulated as part of the first handwritten copies of the *Meditations.*[1] It might be said that they were, from the beginning of the public appearance of Descartes' *Meditations,* part of the corpus and, as such, submitted (together with Descartes' Replies) to the judgment of his other readers. Descartes felt that they were privileged in this way and referred to the objections sent to Mersenne (which we call the Second Objections) as if they were the first ones.[2] Thus it seems he was considering Caterus' objections as part of the *Meditations* themselves. Descartes was very much interested in the objections of "the theologian."[3] As Henri Gouhier noticed, it was as if Descartes had left aside in the Third Meditation some particularities of God.[4] He seized on Caterus' objections as the awaited first opportunity to unfold his thought on this slippery theological question.

Leaving to Professor Verbeek the search for the historical Caterus, I consider here some points useful for understanding the *Objections.* Bannius and Bloemaert were not professional theologians. They asked Caterus to do the job, and their choice was particularly suitable. The archives of the Chapter in Haarlem tell us that Caterus had been required to complete his degree in theology in 1632, on being admitted as canon. In October 1633 the one-year delay he had been allowed was prolonged for one more year. Since we hear no more about it, it is probable that Caterus attended courses in theology in 1633–34 and received his degree

1. To Mersenne, Dec. 24, 1640: "J'ay mis celles [i.e., the objections[de Caterus à la fin, pour montrer le lieu où pourront aussi être les autres, s'il en vient" (AT 3:267).
2. To Mersenne, Jan 28, 1641: "Ce mot n'est que pour vous dire que je n'ay pû encore pour ce voyage vous envoyer ma Réponse aux Objections" (AT 3:292).
3. To Mersenne, March 4, 1641, AT 3:329–30.
4. See Gouhier 1987b, 124, 139–41, 151–52, 155, 158, 163, 164, 171–76.

at Louvain, where the Dutch priests, especially those from Haarlem, were trained. He had already followed part of the curriculum at Louvain in the year 1632. In 1633 the Holy Scripture was taught at Louvain by Cornelius Jansenius, who was to become bishop of Ypres in 1636.[5] (Caterus probably went to live at Sainte Pulchery, Jansenius' former college.) Vopiscus Fortunatus Plempius, one of the professors, had a Jesuit brother, Peter Plempius, who worked in Alkmaar.[6] This Plempius, a specialist in medicine, became rector of the university of 1637. He visited Descartes at Santpoort in August, received three copies of the *Discourse,* and exchanged objections and replies with him. The professors who were giving commentaries on the *Summa* at Louvain at the time were Guillelmus Merchier (1572–1639) and Johannes Wiggers (1571–1639).[7] Merchier taught at Louvain from 1611 to 1639. The only published part of his *Commentary* bears on the third part of the *Summa* (1630); the *Biographie nationale* tells us of a manuscript commentary on the first part, "formerly preserved at the house of the Carmelite Fathers in Antwerp." But fortunately we can read Wiggers' *Commentary* on the first part, which was published posthumously by his nephew Cornelius.[8] Valerius Andreas, in his *Bibliotheca belgica,* gives us a concise biography of Wiggers, who taught at Louvain for twenty-six years: "reliquit Commentaria in totam S. Thomae summam, in quibus nihil nisi summe elaboratum, solidissimis fundamentis subnixum, clarissima et brevi tamen methodo traditium, quidquid in aliorum fisioribus scriptis scitu dignum est." "He left commentaries on the whole *Summa* at St. Thomas, in which there was nothing that was not transmitted with the fullest elaboration, supported on the firmest foundations, with maximum clarity and with brevity although with method: anything whatever that is worthy of record in the written outpourings of others."[9] (This comment had been printed ahead of Wiggers' *De justitia et . . . ,* 1639.) Indeed, Wiggers' theological style is of a rare conciseness. He draws very clear conclusions and omits consideration of the most obvious contributions of Aquinas, in order to write more precisely on the topics usually ignored. His stand is a firm Thomistic reaction against the eclectic synthesis of Suárez.

The First Objections, like the Fourth, are not directed immediately to

5. Orcibal 1989.

6. Adam 1910.

7. Monchamp 1886; on Caterus, see 124–34.

8. Wiggers 1641. Wiggers' commentary was standard reading in the Augustinian circle, and Arnauld used it; it is mentioned by Cambout de Ponchateau in a letter to Arnauld's secretary, Ruth d'Ans (published in Neveu 1969).

9. Andreas 1739.

Descartes: Caterus writes his objections as an answer to his friends Bloe-maert and Bannius. The epistolary fiction is useful; it allows Caterus to choose his field, to select from Descartes' *Meditations* what he wants to comment on, without being compelled to give an exhaustive commentary. Furthermore, the fiction allows him to object in passing to Descartes' mind-body distinction: "I see that I have gone far beyond the normal limits of a letter." [10]

Cleverly, in order to provide a firm basis for his objections, Caterus summarizes Descartes' argument in the Third Meditation. This was the usual procedure in a *disputatio,* and the objector's cleverness consists in his managing the argument and arranging it in the most objectionable way. Somehow, this procedure has often misled commentators who are unaware of the particular rhetorical status of the *Objections.* The practice of the disputation, a basis exercise in any academic training, usually included such a reassessment of the adversary's thesis, arranged in such a way as to make it easier to confute. The fiction of a straw man allows a better stand for the objector; it was also a good pedagogical tool to make sure, by a paraphrase of the opponent's opinion, that the gist of the argument was caught and understood. The *disputatio* was indeed the basic exercise of traditional schooling: first restating in one's own words the opponent's thesis, then confuting it. Thus, the biased quotations of Descartes by Caterus are not the product of intellectual dishonesty; they just belong to the ordinary genre of objections, in an academic disputation. The Adam and Tannery edition is somehow misleading: they use roman type to locate in Caterus' text quotations from Descartes' *Meditations.* But they cheat in the footnotes. Obviously, publication of the *Objections* and *Replies* should return to the 1641 and 1642 format, extending to the *Objections* and *Replies* the minimum requirements of physical accuracy established by Giovanni Crapulli for the *Meditations.* Keeping in mind what we stated about the genre of the *Objections,* we should take the changes and shifts in the quotations into account and handle them very carefully.

In this way of setting his own grounds for combat, Caterus switches from the formulation of the *Discourse—cogito, ergo sum—*to a non-Cartesian formulation, which is to serve as a convenient basis for his own argumentation—*ipsa cogitatio, aut mens sum.* I am thought itself; I am a mind. He continues: "I possess ideas of things and more specially of a very peculiar thing, a very perfect and infinite being." [11] Caterus borrows

10. AT 7:100.
11. AT 7:91–92.

the description of this being from Denys the Areopagite, quoting very loosely from chapter 8 of *On the Divine Names*. This reference was changed, and the text modified, in the French translation of 1647, which refers more accurately to the fifth chapter of Denys' treatise. No literal reference can be found, however, although the description of God as a very perfect and infinite being is a commonplace in Areopagitism. Next Caterus points immediately, *ex abrupto,* to the main thesis of his objections: What is an idea? Are Descartes' proofs of God's existence anything more than the definition of a chimera? What reality can be assigned to the God Descartes points to? Indeed, Caterus' objections are all one and the same: Descartes' God is a strange one. Or, more exactly, Descartes' conception has to undergo a metamorphosis in order to become acceptable to Christian theology.

How can there be such a thing as the cause of an idea? To be a cause means to produce some effect. But no idea can be said to be so real as to be effective, and such a cause as could produce an idea is not acceptable as real: "ergo ideas habeo, causam earum non habeo, tantum abest ut me majorem et infinitum." "Thus I have ideas, [but] I do not have their cause, so much the more do I lack a cause that is greater than me and infinite." [12]

Such a conclusion announces the death of Descartes' God. Its fascination comes from its enunciation in Descartes' own terminology. Indeed, the wheel has gone full circle: Descartes used Scholastic terminology with a biased meaning; his own weapon is turned upon him, and "the learned theologian" can formulate his drastic statements in the very terms of Descartes' exposition. Obviously they do not agree on the meaning of the terms in question. Descartes says that objective being has its place in the mind in the same way that objects usually do: "ex modo quo objecta in illo esse solent," but the expression *esse solent* hides the diffrence.[13] For Caterus, we know the things themselves, and nothing stands between the thought and the object. There is no room, in his theory of knowledge, for such a "thing" as an idea. The way Descartes uses the word *res* might indeed be analyzed as quite interesting: Descartes erects it like a shield hiding the shift from the idea to the object, by calling both of them *res*. Far from being led astray by this conflation, Caterus knows very well what to do about it and does it with great skill: he goes on as if Descartes' meaning were clear, in order to force him into a corner where he can tackle him.

Objecting to himself that one should explain individuals, Caterus an-

12. AT 7:93.
13. AT 7:102, ll. 15, 26.

swers by stating the eternity of essences: "scaphiam scaphiam esse, et nihil aliud." "A boat is a boat and nothing else." [14] The weakness of our minds is the only reason that this seems difficult for us to understand. And Caterus turns back to Descartes' own argumentation, struggling to establish the reality of the cause of the idea of God, which is the main thesis of the Third Meditation. Since Caterus does not share Descartes' definition of causality, he maintains firmly that such a being, if it is at all, can only be a being of reason, *ens rationis,* with no actual existence.

Would the second way Descartes proposes alleviate Caterus' criticism? No doubt it might be more palatable to an authentic Thomist, which Caterus is not. Descartes faces the question of the cause of the *ego,* which is a most peculiar "thing." After giving a summary of Descartes' argument in the Third Meditation *(a quo essem?)* Caterus identifies it: "ille ipsa via est, quam et St. Thomas ingreditur." "It is the very same way that St. Thomas took, the way of the efficient cause, and he borrowed it from Aristotle." (To support the hypothesis that Caterus borrowed his arguments not from Aquinas's *Summa* but from his own Louvain notebooks, we may note that while pretending to quote St. Thomas *(quam vocat),* Caterus uses an expression, *a causalitate causae efficientis,* "from the causality of the efficient cause," unknown to Aquinas, who always says *ex ratione causae efficientis,* "on the basis of the efficient cause." The word *causalitas,* according to the *Index Thomisticus,* does not belong to Thomas' vocabulary.)[15]

Caterus continues, explaining that Aristotle and Thomas did not need to care about the cause of ideas; what they were concerned with was the ultimate causes of actual beings—which ideas are not. Caterus very astutely reiterates the Cartesian *cogito* as he had stated it earlier: "cogito, ergo sum, imo ipsa cogitatio, aut mens sum." [16] I myself am thought or mind.

This leads him to consider the concept of cause as it is linked with *a se esse,* or *aseitas,* "being from itself," or "the state of being from itself." This part of Caterus' objections is very important, and it has not usually been well understood by the commentators. Caterus refers to an opinion of Suárez that he had been taught some years before ("to be taught" is the correct translation of *audivisse.* It conveys the sense of a student's attending a lecture and hearing an opinion; it does not mean that Caterus pretended to be a former student of Suárez himself). Caterus seems to be referring to an opinion of Suárez: A causeless thing is infinite. This opinion is not usually given a reference in Suárez' works; in fact, the textual

14. AT 7:93.   16. AT 7:91.
15. AT 7:48, and 94, l. 15.

reference might be found in the treatise *De divina substantia et ejusque attributis,* with a very good occurrence of *Deus* as *causa sui.*[17] It applies to God, not to the self. But more than that, it is not Suárez' opinion! On the contrary, Suárez quotes this opinion as misleading and argues at length against it:

> [I]n order to confirm the opinion (of God as infinity) by a theological argument, one has to state what is denied to God when he is called infinite (which is taken as an imperfection). It is all the easier since this imperfection has to be removed from God by the highest perfections that we have shown he possesses. Some think this infinity does not deny anything else than God's having no cause of his being *[nihil aliud negari nisi Deum habere causam sui esse],* and they say it is the same thing for God to be infinite and to be *a se* and not *ab alio.* But this is not the common way of conceiving the divine attributes and of talking about them. Nobody, indeed, would understand those two negatives to be formally *[formaliter]* one and the same, that is, to be infinite and to be uncreated or nonproduced; but St. Thomas and all the theologians present those two opinions as distinct divine attributes, and Theodoretus accounts for them as distinct.[18]

Suárez goes on to explain that a distinction of reasons, at least, has to be made: it is more obvious that God is a necessary being without cause than an infinite being. Necessity and the absence of cause can be deduced through the order of causes; from the negation of being from another, you infer the infinity of being. The two negations are not the same, since one is deduced from the other. *Esse a se,* "being from itself," immediately denies dependence on an external cause; *esse infinitum,* "to be infinite," does not deny it *formally.* Since Suárez states here that to be infinite and to be without cause are not one and the same thing, it can hardly be said that he taught what Caterus found in him: the link between being a se and being infinite.

In any case, Caterus does not agree with the statement he assigns to Suárez: like "Thomas and all the theologians"—and like Suárez himself—he asserts that a thing *a se* is not necessarily infinite. It is too bad that Caterus did not know enough Suárez. He might have found in the quoted passage the full development against Descartes' identification between being *a se* and being infinite. But Caterus' ignorance of the whole context tends to confirm the hypothesis that he used his own notes on the lectures he had attended as a student at Louvain a few years earlier. He

17. Suárez 1856–78, 1:46.
18. Ibid.

did not use Suárez' text but limited his argument to the commentary on
the first part that he had had to hear in order to earn his degree.

Quite curiously, Descartes' answer to this argument begins with a biased statement: I never denied, he says, that a thing can be the efficient
cause of itself. (Caterus never claimed that he said so.) While he does not
answer Caterus' argument, he takes advantage of his own statement to
develop what seems to be a needed clarification of *causa sui*.

Caterus' next point bears on a very touchy topic: he allows clear and
distinct things to be true. But what does Descartes mean by saying that
God (the infinite being) is clearly and distinctly known? How can one
know God in such a way? "Infinitum, qua infinitum, est ignotum"; "the
infinite, insofar as it is infinite, is unknown." [19] Descartes had to take this
step: he had to acknowledge God as a clearly conceived being. He finds
himself in direct opposition to St. Thomas, who says, in the well-known
first article of the second question (first part): "[Deum esse] non est per
se notum" at least *quoad nos*, with respect to us. The human mind cannot
understand the Supreme Being, which is beyond its reach and whose concept remains confused. Descartes, in his answer, willingly acknowledges
that he expected such an objection, which readily occurs to anyone. He
then affirms a distinction between understanding and conceiving. The infinite can be conceived without being understood. Caterus' objection
evokes, on Descartes' side, a long and important explanation of what he
means by *infinite* as distinct from *indefinite*.

Caterus continues by granting his statement to his adversary. Let it be
that such a clear and distinct idea of a sovereign being can be conceived.
What could one infer from it? Is it possible to pass from conception to
existence, *a nosse ad esse*? Caterus identifies the Cartesian thinking process with its traditional source: Descartes' argument is close to one of
St. Thomas' objections. Without naming St. Anselm, Aquinas had indeed
objected to the possibilty of an a priori knowledge of God's existence, an
argument close to St. Anselm's famous "proof." Using the traditional way
of the *disputatio*, Caterus proceeds in two steps: first, he identifies Descartes' argument with St. Thomas' objection. Once this is done, he has
only to oppose to it St. Thomas' own answer (and his authority). The
objection—and the Cartesian argument—consist of a logical circle: in
order to infer God's existence *a nosse ad esse*, from knowing to being,
you must first suppose his actual existence. Caterus' example of the existence of a lion is another way of putting the same argument indirectly.

Johannes Wiggers' commentary on the first part of the *Summa* give

19. AT 7:96.

the whole dossier on the second question: "Utrum Deum esse, sit per se notum?" "Whether it be known in itself that God is?" Drawing on Suárez *(Disp. Met. 29) et passim Recentes,* "and the moderns throughout," he quotes St. Thomas, Richard, Hervaeus Natalis, Durandus, the Commentator of Ferrara, Cajetan, and Bannes. He comes to the conclusion: "Deum esse est per se notum secundum se, sive in se, non tamen est per se notum quoad nos sive nobis," "that God is known through himself according to himself, or in himself; however, he is not known in himself with respect to us or by us." This is the traditional Thomist position. To the ancients who said that a natural knowledge of God's existence is offered to our minds, Wiggers opposes the atheists, those of ancient times, but also the moderns "(recentiores) qui honestiori nomine nunc vocantur Politici, qui omni interna fide, cultu et religione Dei abjectis, omnia, etiam ipsam cuiuscumque religionis simulatam speciem, ad solam externam Reipublicae gubernationem, propriumque commodum referunt ac ordinant," "who are nowadays called by the more honest name of Politicals, who, since they had abandoned all internal faith in, adoration of, or reverence for God, turn over and dispose of everything, both private and public, even the simulated appearance of any religion whatsoever, to the sole external governance of the Polity." [20]

This quotation is useful in that it reminds us of the actuality of atheism and of the apologetic character of the contemporary philosophical discussion. But Wiggers' *Commentary,* which is strictly Thomist in its conclusions, also transmits other opinions. Useful for our purpose is Question 15, "De Idaeis," "On Ideas." Wiggers does not rely on the fifteenth question of the first part of the *Summa Theologica:* he uses the *De veritate* (question 2, a. 1), to distinguish between a *forma exemplaris externa* and an internal form:

> externa: qualis est imago, aut scriptura, vel res alia externa oculis obiecta, quae imitanda proponitur;
> interna: quae scilicet mente intrinsecus formatur, ratio nihilominus Idaeae proprie in formam internam, seu mentalem dumtaxat cadit seu quadrat.
> Sed hic inter Scholasticos controvertitur, quomodo Idaeae in mente divina sint constituendae, an objective, an subjective et formaliter, et quidnam proprie sunt ut dicuntur esse in mente divina?
>
> external: such as is an image or a piece of writing or some other external item set before the eyes, which is displayed for the purpose of imitation.

20. Wiggers 1641, *Summa,* pt. 1, question 2.

internal: while the idea is evidently formed intrinsically by the
mind, nevertheless its principle properly falls into or agrees with
the internal or mental form.
But this is disputed among the Scholastics, how Ideas are constituted in the divine mind, whether objectively or subjectively and
formally, and what they are properly as they are said to be in the
divine mind?[21]

Wiggers proceeds to explain that the Scotist tradition (Scotus, Gabriel,
Durandus) takes the ideas for the object itself, the *res objective in divina
mente existente,* "the entity existing objectively in the divine mind." But
others teach, *magis communiter et verius,* "more commonly and more
truly," that the principle of the Ideas does not agree with the objects but
with the divine essence itself. Wiggers' authorities are Cajetan, Alexander
of Hales, Albert, Bonaventura, Aegidius, Capreolis, Henry, and Sylvester
of Ferrara.

As noted above, the *Commentary* of Wiggers was known, according to
Valerius Andreas, as a concise summary of opinions otherwise spread
widely in various treatises. As a matter of fact, Wiggers does not paraphrase the *Summa,* as was usually done in other commentaries. He frequently explains that St. Thomas is sufficiently clear on such and such a
question not to need a commentary, and he limits his teaching to the disputed points. By his treatment of the Scotist position on objective ideas,
he thus provided Caterus with the clue to understanding Descartes' terminology.

Indeed, modern commentaries have too often emphasized that Descartes, by twisting classical Scholastic terms to his own meaning, led his
objectors to strong misunderstandings. As has been noted by R. Dalbiez
and Etienne Gilson, and recently by Marjorie Grene, it was not necessarily so: Descartes' terminology was close to the use it had acquired in the
Scotist tradition, and Caterus was able to identify it as such, according to
what he had been taught at Louvain.[22] Descartes' terminology could not
but induce Caterus (and, indeed, all the Scholastic readers of the text) to
assign him to Scotism (or Ockhamism). The opinions ascribed by Wiggers
to the Scotists were worded in the same way, with expressions similar to
those Descartes uses in the *Meditations.* Apart from any doctrinal rapprochement, the vocabulary involved in Descartes' own metaphysical enterprise leads ineluctably to the comparison with Scotus in the objector's
mind.

21. Wiggers 1641, 134.
22. Dalbiez 1929, 464–72; Gilson 1984, 209; Grene 1991, 12–15.

I shall leave aside the appendix to Caterus' objections constituted by his remark on the mind-body distinction (where he invalidates Descartes' point of view by arguing that Scotus knows of a *formal* distinction that, although it allows us to distinguish between two things, does not allow them really to stand on separate grounds). This last argument stands as an ultimate proof, if necessary, of Caterus' reading of Descartes as a Scotist: he uses an ad hominem argument here by showing how Scotus himself provides him with a *tertium quid* that would permit a distinction without a corresponding existence for the things distinguished.

We could summarize Caterus' objections by saying that they bear on Descartes' definition of *idea;* the problem is that Descartes did not stick to a precise definition. The Cartesian *idea* hides various meanings, and Caterus points out the inconsistency of such an expression as *objective idea*. He thus allows Descartes to revise and enlarge his definition. But the use of the refined tools of Thomist philosophy helps Caterus to set up a network of concepts that serve as a probe for Descartes' argument in the Third Meditation. The ordeal had positive results and enabled Descartes to dig further and deeper in order to reinforce the shaken foundations of his metaphysics. By evoking this result, Caterus affirmed the heuristic power of Thomist questioning. But his objections lead us to go further.

Beyond the classical position of the *Summa,* one may wonder about Caterus' own beliefs. He makes his point by opposing Descartes' argument and shaking what Descartes considered to be his best way of proving God's existence. Although he borrows his main arguments from St. Thomas, one may wonder about his own personal conviction. He is too easily ready to accept Descartes' statement on the truth of what one knows clearly and distinctly to be a hard-core Thomist.[23] One may notice in his arguments the weakness of his own point of view, which is not Thomist enough to be coherent and to stand on its own. More research has to be done on the doctrinal teaching at Louvain, and little more can be said about Caterus until more light has been shed on his student years. In his objections, however, he made the point of raising such decisive arguments against the Cartesian proofs of God that we can wonder how God's existence could any longer be proved. Objections to the *Meditations?* They were intended to be so, but they are more surely objections to God.

23. See AT 7:95, ll. 28–30.

PETER DEAR

4

MERSENNE'S SUGGESTION: CARTESIAN MEDITATION AND THE MATHEMATICAL MODEL OF KNOWLEDGE IN THE SEVENTEENTH CENTURY

Descartes' casting of the central arguments of the *Meditations* in "geometrical" form at the end of the replies to the Second Objections seldom elicits much comment. Descartes agreed to use the geometrical form only at Mersenne's suggestion, and he was openly reluctant to do so.[1] Consequently, commentators interested in Descartes' own thoughts and intentions tend not to focus on these passages; the arguments are sometimes used to elucidate or set in relief the procedures of the *Meditations* themselves, or to flesh out Descartes' views on method, but the "geometrical" structure has attracted little attention in its own right.[2] An examination of the sense in which this "geometrical" exercise counted as such, however, proves to illuminate the central motif of Descartes' book, the idea of meditation itself.

Mersenne's role as an "objector" to the *Meditations* was in large measure one of mediator and communicator. His concern, as with others of Descartes' projects, was to facilitate debate over important issues rather than to challenge the validity of Descartes' arguments.[3] The Second Objections are a more public form of his many letters to Descartes that sought elaboration of difficult points, and the suggestion concerning geometrical exposition accords well with Mersenne's aims. The question of why Mersenne recommended a geometrical rather than a syllogistic deductive structure as the most perspicuous form of argumentation can be answered readily: mathematical demonstration was routinely taken to represent the pinnacle of certain proof, and thus commanded the readiest

I am grateful to Roger Ariew and to two anonymous referees for their helpful comments on this essay.

1. AT 7:160–70 (the layout), 156–57 (the reluctance), 128 (Mersenne's suggestion).

2. E.g., Doney 1978. Risse (1970, 45) mentions the presentation *more geometrico* as a representative of the synthetic rather than the analytic approach, as Descartes himself represents it, but as far as I know, no discussion goes significantly beyond that.

3. For characterizations of Mersenne's philosophical personality, see Lenoble [1943] 1971; and Dear 1988.

acceptance.[4] Descartes himself clearly displays this assumption in his "Dedication" of the *Meditations* to the Sorbonne. At one point, he compares favorably the clarity and conclusiveness of his metaphysical proofs with those of geometry. Descartes' explanation of why, despite this, his "evident" demonstrations are unlikely to receive the immediate, unproblematic acceptance routinely accorded to geometrical theorems is an interesting one. He says: "In geometry everyone has been taught to accept that as a rule nothing is written[5] without there being a conclusive demonstration available; so inexperienced students make the mistake of accepting what is false, in their desire to appear to understand it, more often than they make the mistake of rejecting what is true."[6] In philosophy, however, everyone believes that there is no such thing as an unproblematic demonstration; as a consequence, people are prepared to argue with any philosophical argument whatsoever.[7]

This remarkable piece of sociology of knowledge centers on the statement that nothing is written in a geometrical treatise unless it is conclusively demonstrable. Descartes' remark points toward the significance of sheer textual layout in winning assent for propositions, and more specifically to its significance in rendering the presentation of an argument "geometrical." The most obvious way in which the formal arguments at the end of the Second Replies are "geometrical" is precisely this: they are divided into "Definitions," "Postulates," and "Axioms or Common Notions," followed by "Propositions" together with their "Demonstrations." A "Corollary" even appears for good measure.[8] The appropriate labels, with the appropriate layout on the page, certify Descartes' attempt as a bona fide response to Mersenne's request. No doubt these characteristics alone would have been insufficient to qualify the arguments as "geometrical," but they were clearly quite important: this is a presentation of arguments that can be claimed, with apparent plausibility, to be *more geomet-*

4. For general treatments, see Angelis 1964; Schüling 1969; Risse 1970, esp. chap. 8; and Risse 1964. Jardine (1988, 693–97) discusses the debate in this period on the source of the certainty routinely attributed to mathematical demonstration; his article contains references to the copious literature, especially to a considerable number of studies by G. C. Giacobbe. See also Mancosu 1992. Descartes, of course, argued that (his) metaphysics was more certain even than mathematics.

5. That is, nothing is written, by implication, in a geometrical treatise: Cottingham's translation includes here the terms "proposition" and "in a book"; the relevant phrase is *nihil scribi solere.*

6. AT 7:5; translation slightly modified from Cottingham's, in CSM 2:5.

7. Descartes discusses the same issue in the Second Replies when explaining the relative merits of the synthetic and analytic approaches.

8. AT 7:160–70; CSM 2:113–20.

rico even though it says nothing at all about lines or figures or determinate quantities. Being "geometrical" in the seventeenth century was as much a question of how something was said as it was a question of what it was said about.[9] The issue was not merely one of using axiomatic deductive reasoning; if it had been, appeal to a syllogistic structure of the kind that Aristotle discussed in the *Posterior Analytics* would surely have been a better, and more direct, choice. Labels and layout on the page, "geometry" being the most important label of all, play an evident role here.

The fifth part of the *Discourse on Method* provides a clear example of the way in which the meaning of the term "geometrical demonstration," or its practical synonym "mathematical demonstration," could be used independent of subject matter. Descartes has been discussing the action of the heart, the heating up and rarefaction of blood, the resultant effect on the valves of the heart, the consequent dilation of the arteries, the blood's subsequent cooling and condensation, and so on for the whole circulatory cycle. The account is cast in purely qualitative terms, and yet Descartes clarifies its explanatory virtue through an identification with mathematics:

> In order that those who do not understand the force of mathematical demonstrations and are not accustomed to distinguish true reasons from probable ones do not chance to deny this without examining it,[10] I wish to advise them that this movement I have just explained follows from the mere disposition of the organs of the heart which we can see with the eye, and from the heat which we can feel with the fingers, and from the nature of the blood which we can understand through experience, just as necessarily as does the motion of a clock from the force, situation, and shape of its counterweights and wheels.[11]

Descartes' explanation of the behavior of the cardiovascular system has "the force of mathematical demonstrations" despite the fact that considerations of quantity play no role in it. The foundations of the claimed "demonstration" are not those of geometry or arithmetic; instead, they

9. Reif (1962, 309) notes an increasing tendency among the early seventeenth-century textbook writers whose work she examines to "pattern the format of their manuals on that of a geometrical treatise." This development can also be found in antiquity: Galen recommended the use of mathematico-logical structures in medicine. See Lloyd 1979, 120.

10. Note the close similarity between this sentiment and the remarks in his later "Dedication" of the *Meditations*.

11. AT 6:50; translation adapted from Descartes 1965, 41.

derive from the lessons of the eye, of the sense of touch, and of experience generally.

This example from the *Discourse* illustrates the degree to which the reputation for certainty then attaching to mathematical demonstration rested not only on the clarity of the deductive steps that made up a geometrical theorem but also, and especially, on the perceived self-evidence of the fundamental principles on which the demonstrations rested. In this regard, the technical considerations were the same as those laid down by Aristotle for scientific demonstration generally, including mathematical demonstration.[12] These principles, which in Aristotle's scheme acted as the major premises of syllogistic demonstration, were the primary source of the truth of all conclusions demonstrated from them (contingent statements about particular instances provided the minor premises). The first principles of an axiomatic deductive system are not themselves deductively justifiable, and yet they must be accepted as true in order to ground the derived truth of statements the proofs of which employ them. From a purely practical viewpoint (and the construction of convincing arguments is an eminently practical task), the important thing was to win wholehearted assent from everyone as to their truth. There were no formal deductive rules for demonstrating that truth, by definition: in the *Posterior Analytics* Aristotle talks in this connection about kinds of induction from experience.[13]

Postulates

Descartes appeals to experience, although of a different kind, not just in the example of the heart from the *Discourse* but also in the *Meditations*. Gary Hatfield has argued that the *Meditations* should in large part be taken as an attempt at establishing conviction in the reader (or meditator) of the truth of certain metaphysical axioms, first principles analogous to those of geometry.[14] I want to suggest that this attempt can be seen as an appeal to experience and that it is exhibited most clearly in the formal structure of the geometrical proofs at the end of the Second Replies.

Descartes begins his geometrical exposition by listing the principles on

12. The scientificity (in the Aristotelian sense) of mathematics was contested by some Scholastic philosophers in the sixteenth and seventeenth centuries, although it was defended by others. See Galluzzi 1973; and Crombie 1977. The issue was intimately related to the debate on the certainty of mathematics; see the references in n. 4 above, especially Jardine 1988 and its citations.

13. See the fuller discussion in Dear 1995, chap. 1.

14. Hatfield 1986, esp. sec. 1.

which his arguments are to be founded. The second category of principles, following "Definitions" and preceding "Axioms or Common Notions" is the equally Euclidean "Postulates." In effect, these postulates are instructions to the reader. One should reflect on the unreliability of the senses, on the attributes of one's own mind, on the nature of one's apprehension of self-evident propositions, and on how certain properties follow necessarily from the natures of the things of which they are properties; one should contemplate the nature of God and how he exists necessarily, by his very nature; one should refer back to, and contemplate, the examples Descartes has given in the *Meditations* of clear and distinct perception as well as of obscure and confused perception; and finally, one should think seriously about clear and distinct perceptions until one is convinced that it would be irrational to doubt them.

The literal meaning of the word "postulate" is "request" or "claim." Accordingly, in the French translation of the *Meditations* the Latin word *postulata* appears as *demandes*.[15] In the recent English translation by John Cottingham, a note to this heading reads: "Descartes is here playing on words, since what follows is not a set of postulates in the Euclidian sense, but a number of informal requests."[16] These are indeed informal requests; it is less clear, however, that they are therefore disqualified as postulates "in the Euclidian sense." What, after all, *is* the Euclidean sense of "postulate"?

This is not a question to which there is a straightforward answer; it is difficult to see what all five of Euclid's postulates have in common.[17] The first three concern the possibility of construction, as in the case of drawing a straight line between any two given points. The fourth and fifth, however, are entirely different. The fourth asserts that all right angles are equal (Thomas Heath and others suggest that this functions as a statement of the homogeneity of space), while the fifth is the so-called parallel postulate, which in effect claims that two right lines in the same plane will eventually meet if they are not quite parallel. These two postulates assert properties themselves, rather than the existence of those things of which the properties are supposed. In a sense, Euclid simply asks his readers to accept certain propositions as a precondition for enabling the work that follows.[18]

Euclid's formulation seems largely to have been maintained in the chief

15. AT 9:125.
16. CSM 2:114n.
17. See Euclid 1956, 1:119–20, 195. Lloyd 1979, chap. 2, is very illuminating on the whole issue of the foundations of Greek deductive systems.
18. Euclid 1956, 195–220.

editions of the *Elements* that were produced in the first century or so of printing.[19] By contrast, the commentary version of Euclid's *Elements* that was designed for and used in Jesuit colleges in the late sixteenth and seventeenth centuries deals with this issue in a slightly different way. The work is by the Jesuit pedagogue Christopher Clavius; its ubiquity guarantees that Descartes used it at La Flèche.[20] In the "Prolegomena" prefixed to the commentary, Clavius explains the various kinds of "principles" used by mathematicians as the foundation for their demonstrations.[21] Following "definitions," a methodologically unproblematic category in Clavius' treatment, he presents "postulates" and "axioms":

> The second kind [of principles] comprises *Petitiones,* or *Postulata,* which indeed are very clear and perspicuous in the science that is under consideration, so that they need no confirmation, but merely demand the assent of the hearer, and neither is there any hesitation or difficulty in explaining [them]. To the third kind are referred *Axiomata,* or common notions of the mind, which are so manifest and evident not only in the science in question but also in all others that he can dissent from them with no cause who will rightly understand the very words.[22]

Clavius thus seems to distinguish between "postulates" as evident principles proper to a particular science and "axioms" as evident principles

19. E.g., Campanus de Novarra 1482, 2nd page. Campanus' postulates are not identical to Euclid's; he combines Euclid's first two and adds one, perhaps implied by the first (and presented by Pappus as an axiom: see Euclid 1956, 195–96, 232), to the effect that two straight lines cannot enclose a surface. Commandino gives Euclid's five in his Latin and Italian versions, of which *Elementorum Euclidis libri tredecim* (1620), which I use, is based upon his Latin translation (see Euclid 1956, 102). The edition of the Italian version that I have used is Commandino 1619.

20. Clavius 1611–12, vol. 1, "Commentaria in Euclidis Elementa Geometrica." For a general synthetic account of Jesuit education in Descartes' time, see Shea 1991, 4–6. Hervey (1952, 78) quotes Pell to Cavendish, March 2/12, 1645/46, on a conversation with Descartes: "He says he had no other instructor for Algebra than ye reading of Clavy Algebra above 30 yeares agoe"—precisely when Descartes was at La Flèche. Cf. AT 10:156, 156n, on Descartes' use of Clavius' algebraic terminology in 1619.

21. The Prolegomena appeared with the commentary in all editions from 1574 onward, including *Opera mathematica.*

22. Clavius 1611–12, vol. 1, "Prolegomena," 9: "Secundum genus complectitur Petitiones, sive Postulata, quae quidem adeo clara sunt, & perspicua in illa scientia, quae in manibus habetur ut nulla indigeant confirmatione, sed auditoris duntaxat assensum exposcant, ne ulla sit in demonstrando haesitatio, aut difficultas. Ad tertium genus referuntur Axiomata, seu communes animi notiones, quae non solum in scientia proposita, sed etiam in omnibus alijs ita manifesta sunt & evidentia, ut ab eis nulla ratione dissentire queat is, qui ipsa vocabula recte perceperit."

common to all sciences. The matter becomes less clear, however, when
one examines the items listed under the heading "Postulates" in Clavius's
Euclidean geometry itself. First, Clavius does not list Euclid's five.[23] The
parallel postulate appears instead as Axiom 13, with the remark that
Clavius will attempt to demonstrate it as a scholium to Proposition 28—
a demonstration founded on Definition 34.[24] The latter is genuinely Euclidean, a definition of parallel lines as right lines in the same plane that,
even when produced to infinity, never meet.[25] Clavius also moves to his
axiom section Euclid's curious postulate concerning the equality of all
right angles, where it appears as Axiom 12.[26] The four postulates that
Clavius presents seem to fit more closely the structural analogy with Aristotle's category of "hypotheses" than did the original five: they are all in
one way or another statements of existence or constructibility. Clavius
retains Euclid's first three. These concern the constructibility of a straight
line between any two given points; the indefinite producibility of a
straight line in either direction; and the constructibility of a circle of any
given radius around any point. He adds to these a fourth, also having to
do with constructible existence, which states that one can always take a
magnitude greater or less than any given magnitude.[27]

After having listed these definitions, postulates, and axioms, Clavius
acknowledges that he has used a particular understanding of these categories, one given by Proclus and championed by the Greek geometer
Geminus, that actually differs from that underlying the divergent account
of the Prolegomena (also derived from Proclus).[28] Axioms and postulates
are now differentiated by defining axioms as unproven propositions the
truth of which is easily grasped by the understanding, such as "the whole
is greater than its proper part," and postulates as unproved propositions
that assert the possibility of *doing* something—that is, of performing con-
structions. Proclus himself had pointed out that if this characterization is

23. Clavius 1611–12, vol. 1, "Commentaria," 22–23.

24. Ibid., Axiom, p. 25; Prop. 28, pp. 48–53.

25. Clavius' Definition 34 corresponds to Definition 23 in the Greek Euclid (Euclid 1956,
190–94); it is the twenty-second in the first printed edition by Campanus (Campanus de
Novarra 1482, 2d page); Commandino (1620, 4) gives it as Definition 35. Clavius uses it
only for his "proof" of the postulate as part of Proposition 28. See Maierù 1978, 1982,
1989. Naux (1983, 333–34) discusses Clavius' handling of the "theory of parallels," indicating its novelty. Clavius' innovativeness in his treatment of parallels is also mentioned in
Euclid 1956, 194.

26. Clavius 1611–12, vol. 1, "Commentaria," 25.

27. Attributed to Pappus as an axiom; see Euclid 1956, 232.

28. Clavius 1611–12, vol. 1, "Commentaria," 26.

accepted, Euclid's fourth and fifth postulates will not count as true postulates because they have nothing to do with constructibility.[29] In his practical classification, Clavius thus accepts Proclus' suggestion, removing Euclid's final two postulates and putting them under "Axioms," a heading for which they now qualify. Clavius' new postulate is, like the first three, about constructibility.

In geometry, an assertion of constructibility is also an assertion of potential existence, because it states the possibility of actualizing that existence through construction. When Clavius uses constructibility as the characteristic feature of his postulates, he is asserting, in effect, that the things of which he speaks are real. The Latin word *suppositio,* used to translate Aristotle's *hypothesis,* allows constructive "postulates" to be understood as equivalent to a category of "mathematical suppositions." [30] On that basis, they can usefully be compared to the "suppositions" of empirical sciences. An empirical supposition was typically established, in Scholastic-Aristotelian methodology, by appeal to uncontested experience, and once that was done, it could be used as raw material—a premise—in constructing a scientific, causal demonstration.[31] Ideally, its truth was already established through common experience: everyone had the same personal experience on tap to which appeal could be made in establishing the empirical principles of the science.[32]

29. Proclus 1970, 140–43.

30. Aristotle himself characterizes *hypotheses* as statements of existence in *Posterior Analytics* 1.10; this corresponds closely to the assertions of constructibility that Clavius presents as his postulates (see Lloyd 1979, chap. 2). The picture is slightly confused by Aristotle's additional distinction between "suppositions" (*hypotheses*) and "postulates." See Euclid 1956, 117–22. The issue of constructibility is well known to historians of mathematics as one of some complexity, since opinions differed among Greek and early modern mathematicians over the instrumentalities permissible in the construction of curves. The point is particularly sharp in regard to so-called "mechanical" curves, which Greek mathematicians routinely declared out of bounds but which Descartes argued should be allowed as legitimate (see n. 46, below). For the purposes of the present argument, however, these particularities will be ignored; the broad epistemic significance of constructibility seems the same among the differing positions.

31. The term "supposition" was subject to uses of varying strictness; for its seventeenth-century understanding in this quite commonplace sense, see Dear 1987, sec. 3; Dear 1995, chap. 2; with more focus on Aristotle's own usage, Wallace 1981; and, on variant meanings in Aristotle, Wallace 1992, 142–44. Aristotle (*Posterior Analytics* 1.10) has it that a "supposition" (*hypothesis*) is a statement that the learner is prepared to accept even though it is not proved; transferred to empirical matters, this could be seen as tantamount to an acceptance of an empirical assertion as part of general knowledge.

32. This ideal was seldom realized, of course, as the practice of various sciences made clear. Cf. Dear 1987.

Constructibility

The relevant meaning of the word "experience" itself depended on the goal of achieving scientific, deductive demonstration. It did not mean immediate perception; Aristotle says that "one necessarily perceives an individual at a place and at a time, and it is impossible to perceive what is universal and holds in every case." The point, as Aristotle goes on to say, is that "since demonstrations are universal, and it is not possible to perceive these, it is evident that it is not possible to understand through perception."[33] Instead, "experience," according to Aristotle, should be defined in this way: "Memories that are many in number form a single experience."[34] This idea underpins Aristotle's central concept of induction as it is rooted in sensory evidence.[35]

In the case of the mathematical suppositions called "postulates," especially in Clavius' Aristotelian reading of Euclid, appeal is made to the practical "experience" of geometrical construction. The experiential knowledge that something can by its nature be done, and that therefore the outcome is always possible, serves as an experiential universal, fit for scientific service.

However, mathematical or geometrical reasoning was not always seen in this period as founded on experience. It could sometimes appear as something quite opposed to experience, if that word is taken in its standard Aristotelian sensory connotation. Recall Galileo's famous celebration of Copernicus, in which he praises the astronomer for having had the courage to abandon the evidence of his senses and assert, on purely mathematical grounds, the motion of the earth.[36] A similar attitude, expressed even more clearly, appears in Montaigne: "I have been told that in geometry (which thinks to have gained the highest point of certitude among the sciences) there are found inevitable demonstrations subverting the truth of experience. Jacques Peletier told me while staying with me, that he had found two lines approaching one toward the other so as to join up, that he verified nonetheless could never meet, even extended to infinity."[37]

33. Aristotle, *Posterior Analytics* 1.31; translated in Aristotle 1984, vol. 1.

34. Ibid., 2.19 (note typographical error: "from" for "form").

35. For examinations of the varying senses of "induction" in Aristotle and in the medieval and Renaissance periods, see Pérez-Ramos 1988, chaps. 15, 16; and Milton 1987, which in part criticizes the well-known discussion in Hacking 1975. See also Wallace 1992, 165–70, for more on the Zabarellan sense of the term.

36. See Galileo 1953, 328.

37. Montaigne, *Essais,* bk. 2, chap. 12.

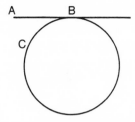

Figure 4.1

Montaigne's remarks lead us back again to Clavius. The example Montaigne takes from Peletier indicates an attitude toward mathematical truths that was also at the core of another more controversial position adopted by Peletier, one fiercely opposed by Clavius in his Euclid commentary. Peletier was a leading French mathematician of the second half of the sixteenth century; in a Euclid commentary of his own, published in 1557, he maintained that so-called contact angles do not constitute quantities. A contact angle in its simplest form is the displacement between the circumference of a circle and a tangent to that circle, measured from their point of contact, just as an ordinary rectilinear angle is a rotatory displacement between two right lines (see figure 4.1; the contact angle is represented by ABC). Peletier's reason for rejecting the proposition that a contact angle is a quantity relied first of all on the argument that contact angles and rectilinear angles are homogeneous magnitudes *(corrationales)* because they are produced in the same way *(congeneres)*. Once granted, this point allowed Peletier the further argument that the contact angle was of no quantity at all—as its now legitimated comparison with rectilinear angles showed. At the point of contact, the straight line and the infinitesimal portion of the circle's circumference are parallel to one another and thus subtend no angular quantity whatsoever. To put it another way (that recalls the argument that so impressed Montaigne), as the two lines approach the point of contact, the angle (of the ordinary kind) between the tangent and the adjacent segment of the circle decreases to zero. Euclid had expressed it, in *Elements,* book 2, Proposition 16, by saying that the contact angle is less than any acute rectilinear angle. Peletier, observing that a contact angle cannot therefore be divided or augmented, claimed that a contact angle cannot be a quantity, since a quantity is by definition something that can be divided or augmented.[38]

38. Peletarius 1557, 73–78. For the most forceful statement of the argument that contact angles and rectilinear angles are homogeneous magnitudes, see the condescending syllogistic exposition of the point in Peletarius 1579, 7v.

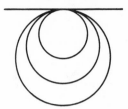

Figure 4.2

In 1563 Peletier reiterated the argument in another publication. Clavius vehemently disagreed with Peletier's position, taking Peletier to task in the first edition of his Euclid in 1574.[39] Peletier unleashed a scorching rebuttal in 1579, and Clavius responded at inordinate length in subsequent editions of his textbook.[40] In his piece of 1579, Peletier endeavors to counter the intuition that the contact angle must be a quantity, insofar as one can always draw new circles of greater or lesser radius tangent to the same point, the circumferences of which therefore pass inside or outside that of the first circle (see figure 4.2). All such circles, he observes, are similar figures, and therefore the actual contact angle between the tangent line and each of the circles, whatever its radius, must always be the same—that is, no quantity at all.[41]

Clavius' central argument is that there is indeed a determinate quantity, an angle of a particular kind, between the tangent and the circumference, but that it cannot properly be compared with acute rectilinear angles as Peletier claimed. It is a different kind of magnitude, but a magnitude nonetheless, because it can be *measured*. The contact angle is a quantity because it can be divided—not by a straight line, certainly, which will always give an angle of zero because it will always have to coincide with the tangent itself, but by another circle. Using the same diagram as the one employed by Peletier (a version of figure 4.2), Clavius points out that the act of drawing a larger circle outside the original circle clearly creates a smaller contact angle between circle and tangent. The circumference of the larger circle divides the contact angle formed by the smaller, and thus

39. See Maierù 1990. Maierù 1984 gives useful background. Popkin (1979, 98 and n. 48) mentions both Leonard Marandé and David Hume as commentators on this dispute who saw it as a conflict between the information of the senses and mathematical reasoning.

40. Peletarius 1579; Clavius 1611–12, vol. 1, "Commentaria," 117–26.

41. Peletarius 1579, sec. 1. Peletier (4r) stresses that the point of his argument is to show, not that a contact angle is nothing (*nihil esse*), as he says Clavius claimed him to have said, but only that it is not a quantity (*quantitatem non esse dixi*); see also 7v.

supports the claim that these angles are quantities.[42] The fact that the dividing line is not straight is irrelevant to Clavius: he has shown that a clear operational definition can be given of the concept of "quantity" in the case of contact angles. They are quantities simply because constructional operations can be defined that give meaning to the claim that they can be diminished or augmented indefinitely.

The central point that emerges from this exchange is that Clavius' approach to the geometrical puzzle, in contrast to Peletier's, is fundamentally constructivist. Perhaps Peletier's ought to be called "Platonic," to point up the contrast, but certainly Clavius, despite his frequent bows to Proclus as a philosophical vindicator of mathematicians, remains true to the Aristotelian principles that are apparent in his handling of the geometrical postulates at the beginning of his Euclid edition. Geometry remains true to experience because mathematics itself is concerned with the abstracted quantitative attributes of real things in the external world.[43] Only operational, constructivist accounts of geometrical figures and concepts are philosophically admissible, as Clavius' geometrical postulates make clear.

Curiously, in 1662 Antoine Arnauld and Pierre Nicole, the authors of the Port-Royal *Logic,* mentioned the Peletier-Clavius dispute as an example of how even geometers sometimes fall afoul of the tendency to confuse words with things themselves: "Who doesn't see that the whole dispute could be ended in one word, by each asking the other what he understood by the word 'angle'?"[44] But this judgment is too facile. The dispute was not over the word "angle"; it concerned the meaning of "quantity." Those who attempted to resolve it appealed to the epistemological foundations of mathematical knowledge.

For Clavius, reality, in the case of geometrical postulates, meant constructibility: real essence was equivalent to potential existence, perhaps.[45] Descartes' own use of mechanical constructibility in the definition of mathematical curves is well known;[46] it went along with his use of a kind of "proportional compass" to establish a new concept of general magni-

42. Clavius 1611–12, vol. 1, "Commentaria," 119.

43. Aristotle, *Metaphysics* 11.3 (1061a28ff).

44. Arnauld 1662, 1:319. Isaac Barrow also regarded the matter as one of adequate definition in his "Lectiones mathematicae XXIII" (delivered in the early 1660s, originally published in 1683); see Barrow 1860, 178 (Lect. XI).

45. This was a position held by the late sixteenth-century Jesuit metaphysician Francisco Suárez; see Dear 1988, 54.

46. Molland 1976; Bos 1981; Lenoir 1979. Funkenstein (1986, 315–16) sets the point within a wider discussion of "knowing by doing" in the seventeenth century, for which Hobbes can stand as the exemplar. Gaukroger 1992 stresses a different viewpoint.

tude that would exclude dimensionality from the operation of multiplication applied to geometrical magnitudes.[47] Founding geometrical objects in the nature of their constructibility, or founding geometrical postulates in the potential *experience* of constructibility, were central parts of Descartes' understanding of geometry.

It should therefore be entirely unsurprising that in the "postulates" to his geometrical argument in the Second Replies Descartes does not ask his readers to grant him certain propositions. Instead, he asks that they engage in certain reflections, concerning the senses, the nature of self-evidence, God's existence, and so on. In effect, Descartes requires his readers to go through the *Meditations* themselves all over again; it appears that the real content of the *Meditations* is packed into the "postulates" in outline form. For Descartes, the formal geometrical structure was really beside the point, as he explained to Mersenne, because it hid all the real work.[48] The real work lay in rendering acceptable the fundamental principles on which the deductive structure was built. To put it another way, and recalling Gary Hatfield's observations, in his "postulates" Descartes simply codifies and puts in their appropriate category the very principles that the *Meditations* were designed to establish. Their designation as "postulates" is not, as Cottingham put it, a "play on words," because just as postulates in geometry are (on the constructivist interpretation found in Proclus) statements of the constructibility of certain lines or figures, so these *metaphysical* postulates are statements of the conviction that a meditator can develop of the truth of certain basic propositions. They assert that the faithful reader may actualize a potential conviction through some form of consideration or contemplation. And Descartes even includes tips on how to do it.

This potential experience of conviction was to be realized through a process that, by its very nature, could not be codified in formal rules. The words that Descartes uses for the procedure are ones such as "contemplate," "consider," "examine," and their cognates.[49] The question that therefore arises is this: How different are these words from the word "meditate" itself? How would Descartes, or a contemporary of his in the same cultural milieu, have associated them with each other and with particular mental procedures?

47. See Schuster 1980, esp. 47–51.
48. AT 7:156.
49. Relevant words from the section "Postulata" (AT 7:162–64) include *considerare, contemplandâ;* Descartes asks that readers *advertant, examinent,* their minds and various mental objects.

Meditation

One association of the word "meditation" is, of course, well known as a way of making sense of Descartes' enterprise: the association with devotional meditation or "spiritual exercises." A number of scholars have attempted to illuminate the character and strategy of Descartes' *Meditations* by comparing it with texts such as Loyola's *Spiritual Exercises,* St. Augustine's *Confessions,* and diverse other religious models that were familiar in Descartes' world.[50] There can be no question of the strong and important parallels in play here.[51] But the religious connotation of the word "meditation" was not the only one operative in this period.

A work by the French Jesuit Honoré Fabri, a textbook of philosophy published in 1646, contains an extended introductory treatment of "method" in various sciences, including logic, physics, and ethics.[52] The term "method" is used quite vaguely, apparently to refer to any kind of procedure or way of going about things. One of the early sections, on method in the sciences in general, carries the heading "De methodo meditationis," "On the Method of Meditation."[53]

Fabri does not provide a formal definition of this particular "method" of philosophizing, but he presents it as one of a set of "methods" specifically designed to assist students—"ad studendi, discendique." His overall methodical concern, then, is with ways of studying and learning. There are four such techniques in all: *lectio, meditatio, exercitatio,* and *scriptio* (reading, meditation, practice, and writing), of which all but *meditatio* are fairly self-explanatory.[54] Fabri's notion of "meditation" in the present context amounts to thinking things over in a disciplined way, so as to

50. Hatfield 1986, sec. 1; Stohrer 1979 (with many earlier references). Rubidge (1990) argues forcefully that Descartes' apparent appeal to the genre of devotional meditations, as discussed with differing emphases by, among others, Hatfield and Stohrer, lacks the specificity of reference that those authors claim. Instead, Rubidge sees no real philosophical content to Descartes' borrowing of the mantle of "meditations" in this devotional sense. I argue, however, that another contemporary sense of "meditation" did indeed provide philosophical meaning to Descartes' enterprise and that Hatfield's characterization, whether or not it corresponds to Augustinian devotional meditation, accurately captures that meaning.

51. Even Rubidge acknowledges this, while questioning its philosophical, as opposed to its rhetorical and propagandistic, meaning; see Rubidge 1990, esp. 48–49.

52. [Fabri] 1646. The work is accepted as Fabri's own, despite the title's attribution.

53. Ibid., 23–24.

54. Ibid., 23, with back-reference to 21.

understand them properly; but the nature of the discipline remains unclear. This fairly casual use of the word seems to have been quite common; for example, Clavius, in the preface to his Euclid commentary, says: "We have added in many places various problems and theorems considered not unpleasant, and not alien to the scope of geometry, which in part we have drawn from Proclus, Campanus, and other authors, in part from our own exertions, and by unremitting meditations we have completed them."[55] Meditation was purposeful thinking.

As with his other categories, Fabri provides under the heading "Methodus meditationis" what amount to study tips—in this case, pointers as to how to think about things. Of the twenty he lists, a few will suffice to provide an indication of their character. Most are instructions of some kind, usually just a short sentence. The most common phrases in them are of this sort: "Examine diligently," "Interrogate yourself," "Deliberate," "Consider attentively."[56] Precepts of particular interest for present purposes include the first: "Examine diligently the primary reason for the opinion that you consider, because, that standing, although other reasons might fall to the ground, that opinion will stand," and the second: "Interrogate yourself frequently as to whether what you are thinking is true, and why."[57] The most interesting of these precepts on meditating, however, come at the end of the list:

> Seventeenth. You learn to separate in your mind all causes, to segregate each one from the others, to consider them all separately; that is, so that everything, as if stripped of its coverings, appears more distinctly.
> Eighteenth. Consider attentively whatever you take as a supposition, since it commonly happens, that from a mistaken supposition not having been anticipated, every error arises; in this regard, it will be of benefit to examine each one [i.e., supposition] separately. . . .
> Twentieth. For the rest, to philosophize is to meditate rightly:

55. Clavius 1611–12, vol. 1, "Commentaria," 10: ". . . adiunximus multis in locis varia problemata, ac theoremata scitu non iniucunda, neque à scopo Geometriae aliena, quae partim ex Proclo, Campano, aliisque auctoribus decerpsimus, partim proprio (ut aiunt) Marte, assiduisque meditationibus ipsi confecimus."
56. [Fabri] 1646, 23–24: *diligenter examina; interroga te ipsum; perpende; considera attentè.*
57. Ibid., 23: "Primum. Primariam rationem illius sententiae, quam meditaris, diligenter examina, quippe illâ stante, etiamsi aliae ruant, ipsa sententia stabit.
"Secundum. Interroga te ipsum frequenter, an verum sit quod cogitas, & quare."

moreover, indeed, to have without interruption the abstracted and estranged mind of a sleeper.[58]

Many of these interesting tips to help budding philosophers think properly, to "meditate" rightly, have a general resemblance to Aristotle's advice in the *Topics*. Their presentation, however, appears unsystematic. It forms no part of any organized discussion of dialectical procedure, nor, indeed, does it contain any reference to Aristotle. Instead, Fabri's pedagogical authority is an Italian Jesuit, Antonio Casiglio, or Casilius. Casilius had written an *Introduction to Aristotle's Logic*, which first appeared in 1629. The fourth chapter of its introductory section is called "De ratione studendi"; it is the source of Fabri's four categories of *lectio, meditatio, exercitatio,* and *scriptio*.[59] There is, however, one slight difference in terminology. Fabri gives close paraphrases of the contents of Casilius' brief chapter, presenting in addition his own lengthy expansions and additions, but the category for which he uses the label *meditatio* Casilius himself had designated with the word *speculatio*. The terms for the other three categories are identical.

Casilius warns his reader that "speculation," by its nature, cannot be reduced to firm rules, because different minds work in different ways. He thus leaves no doubt about the status of the five instructions that follow: they are hints, tips, or suggestions, not logical procedures. Nonetheless, they represent the true nature of "speculation" or "meditation" in this context. The first advises that

> after one has directed himself to grasp the author's opinion well, including the objections of others and their major reasons, he should then begin to judge carefully whether the individual propositions of the arguments are true, by resolving them as much as possible to the most firmly known principles; and whenever they cannot be resolved, or are not consistent with the others, that will be a sign that the argument produced from such propositions is infirm.[60]

58. Ibid., 24: "Decimum septimum, Disces rationes omnes, mente separare, unam ab aliis segregare, seorsim omnes considerare, ut scilicet res omnis suis veluti corticibus exuta, distinctiùs appareat.

"Decimum octavum. Considera attentè, quid supponas, quippe communiter accidit, ut ex falsâ suppositione non praevisâ, omnis error nascatur; ad hoc proderit singula seorsim examinare. . . .

"Vigesimum, Caeterùm Philosophi [reading as *Philosophari*] est rectè meditari; at verò somniatoris mentem abstractam & veluti abalienatam continuò habere."

59. I use the third edition: Casilius 1643, 19–21.

60. Ibid., 20: "postquam advertit quis, se benè tenere Auctoris sententiam, ut etiam obiectiones aliorum cum suis rationum momentis: tunc incipiat perpendere singulas argu-

The second instruction is a warning to the student not to follow the opinions of the professor too slavishly; it asserts the importance of investigating things themselves rather than accepting the judgment of another and the importance of clearing away the ambiguity of propositions and terms in arriving at the truth. The third advises confused students to vocalize their thoughts, so as to help develop mental clarity, and to use plenty of examples. The fourth tells the student to separate things that are certain from those that are uncertain and to avoid the extraneous. Finally, the fifth recommends the use of the syllogism, the most effective tool in refutation or confirmation.[61]

Many of these study tips, of course, call to mind Cartesian recommendations on proper meditation. They suggest that some of the connotations of words such as "meditation" in Descartes' world involved ideas of informal reasoning—perhaps it could be called dialectical reasoning, although it is not dignified with so precise a term. Part of what was at stake in these informal, and in fact strictly uncodifiable, procedures was the establishment of conviction regarding propositions. Those propositions might be reducible, given enough hard thought, to basic principles that were *per se nota*—a procedure that recalls Pascal's discussions in his "De l'esprit géometrique."[62] But the word "meditation" could also apply to procedures whereby the thinker becomes convinced of the truth of the basic principles themselves. This is precisely the issue that underlay Descartes' observation, in his dedication of the *Meditations* to the Sorbonne, of how much more difficult it is to persuade people to accept basic, true principles in philosophy than it is in geometry.[63]

"Meditation" of this kind played a role in the demonstrative regress, best known through Zabarella's influential discussions of it at the end of the sixteenth century. The basic structure was as follows: One starts with a phenomenon to be explained. Through a process of "resolution" or logical analysis one determines a candidate cause for that phenomenon. Subsequently, a process of "composition," or synthesis, is developed whereby the phenomenon is logically deduced from the putative cause

---

mentorum propositiones, an verae sint, resolvendo quantum potest illas ad principia notissima; & cum resolvi non possunt, vel non cohaerent cum alijs; signum erit argumentum conflatum ex talibus propositionibus esse infirmum."

61. Ibid., 20–21.

62. Pascal 1963.

63. Hatfield (1986, 49 and n. 9) notes that François de Sales, writing on spiritual exercises, asserted that the term "meditation" ought properly to be reserved for the discursive, argumentative parts of the exercises. Even here, then, the term had a dialectical cast.

and hence causally explained in the regular way. The difficulty lies in the move from the resolutive stage to the compositive stage.

The aim of the resolution is to display some factor in the situation that is analytically associated with the phenomenon. But that factor cannot be used as the basis for a subsequent deductive causal *explanation:* the resolution shows only that the factor is a constant concomitant of the phenomenon; it cannot, by its logical nature, show that the factor is a *cause* of the phenomenon. As an example, consider the phenomenon of cold weather in winter. A constant concomitant of that situation is the prominence (in the Northern Hemisphere) of the constellation Orion, in contrast to its poor visibility during the summer. Nonetheless, one might judge it implausible that the cause of coldness in winter is Orion's visibility. By contrast, another constant concomitant, the much lower altitude of the sun in the sky, might well appear as a plausible potential cause. Some independent determination, by its nature not readily formalizable, has therefore to be carried out in order to connect the resolutive and compositive stages.[64]

Zabarella and others called the requisite procedure *negotiatio;* the terms *consideratio* and *meditatio* were also sometimes used.[65] It was an indeterminate process of "thinking things over" aimed at creating the conviction that there was indeed a necessary causal relation between the phenomenon and its concomitant, thereby allowing the construction of a causal demonstration. Although Zabarella's justification for it relied on particular metaphysical assumptions of his own, once again, as in Casilius and Fabri, informal reasoning appears as a perfectly valid way of establishing conviction of the truth, even (or perhaps especially) in the case of fundamental, necessarily indemonstrable principles—"suppositions" or "postulates." And the word *meditatio* legitimately applied to that kind of intellectual exercise.[66]

From this perspective, then, the "Postulates" section of Descartes' "geometrical" argument in the *Meditations* represents the establishment of fundamental principles through a process of "meditation" in just this sense—precisely the sense that it carried in common philosophical usage.

Descartes' biographer Adrien Baillet tells how Descartes, while a pupil at La Flèche, was routinely allowed to get up late in the morning, purportedly on account of his weak health, but also because Father Charlet, who

---

64. For a lucid treatment, see Jardine 1988, esp. 686–93; and 1976, esp. 280–303.
65. See Jardine 1988, 687.
66. See above, at n. 55, for Clavius' use of the word.

had personal responsibility for him, "noticed in him a mind given naturally to meditation."[67] Indeed, Descartes considered throughout his life
that he thought best in the mornings, while lying in bed. Now consider
the nineteenth item in the Jesuit Fabri's list of tips for the implementation
of the *methodus meditationis:* "Some people speculate best by day, in the
light; others at night, in the dark; others standing; others sitting or walking, or even lying down. Everyone knows his special aptitude; let him
employ it."[68]

Descartes knew very well how to meditate in philosophy; he preferred
lying down, and it was not a procedure lacking pedagogical authority.
The *Meditations,* unusually, exhibited it outside the classroom, but when
Mersenne indicated his preference for a more conventional form of public
presentation, a mathematical one, Descartes had to put this crucial procedure into the category that most nearly appealed to it: that of geometrical postulates.

67. Baillet 1691, vol. 1, 18: "[I]l remarquait en lui un esprit porté naturellement à la médi-
tation."

68. [Fabri] 1646, 24: "Aliqui diu in luce, alij noctu in tenebris, alij stantes, sedentes alij,
vel ambulantes, vel etiam decumbentes meliùs speculantur; quisque suum genium norit,
eo utatur."

— ✤ 5 ✤ —

J.-B. MORIN AND THE SECOND OBJECTIONS

Of the seven sets of objections to the *Meditations,* two stand out as being a bit different, the Second and the Sixth. In every other case we can identify one person, a philosopher or a theologian, who is the author of those objections. In the case of the Second and the Sixth, though, we are dealing with objections that have been collected by one person, the ever-present Father Mersenne, but that purport to represent the work of a number of other scholars, who remain unidentified. For most purely philosophical purposes, this does not matter a great deal; after all, an idea is an idea, whoever happens to have it, and if what is important is just the confrontation of ideas with one another, then the particular identity of the authors in question, those who contributed to these two sets of objections, is relatively unimportant.

But for those of us with a more historical approach to the texts, this is an unfortunate gap. First of all, it is intrinsically interesting from a historical point of view to know who may have contributed to the drafting of these objections. But more important, in order to understand the objections, their meaning and import, it is very important to know something about their authors. In particular, I shall argue that, behind the scenes in the Second Objections and Replies, there is not merely an author but a text that is important for understanding what Descartes is doing, a text that it is implicitly referred to in the Second Objections and is the direct object of Descartes' reply in the geometrical presentation of the arguments that follows the Second Replies.

The author in question, I claim, is Jean-Baptiste Morin, astrologer, physician, and professor of mathematics at the Collège Royal, and the text in question, his *Quod Deus sit,* a small tract, published in Paris in 1635, in which he presents an argument for the existence of God in geometrical

Work for this paper was supported by the National Endowment for the Humanities, an independent Federal Agency, under grant RH-20947 and by the National Science Foundation under grant DIR-9011998. I would like to express my sincere thanks to those agencies for their kind support.

form, with definitions, axioms, and a string of theorems. I will begin with a brief biographical sketch of Morin, one of the more curious savants of his time. Then I will discuss his relations with Mersenne and Descartes and make the case that he was behind at least certain portions of the Second Objections to the *Meditations*. Finally, I will discuss Descartes' reaction to Morin's pamphlet, and Morin's later reaction to Descartes, concentrating in both cases on the question of the geometrical presentation of metaphysics, Morin's advocacy, and Descartes' critique.

Jean-Baptiste Morin

Jean-Baptiste Morin was born on February 23, 1583, in Villefranche-en-Beaujolais, which made him just a bit more than thirteen years Descartes' senior.[1] Morin's early years were not easy; illness and the necessity of earning his keep prevented him from pursuing the studies in natural philosophy that interested him from his earliest years. (One interesting event is a grave injury he suffered in 1605, at the age of twenty-two, *mulieris causa,* causing him to flee Villefranche. Though he never married, Morin seemed always to have had a weakness for the ladies.) Finally, in 1608, at the advanced age of twenty-five, he obtained the protection of Guillaume du Vair, Premier-Président of the Parlement d'Aix, who enabled him to have lessons in mathematics, then helped him resume his studies, first in philosophy, then in medicine. Morin graduated from Avignon in May 1613. Shortly thereafter he went to Paris, where he entered the service of the bishop of Boulogne, Claude Dormy, as physician. Dormy encouraged Morin's studies of astrology and alchemy, and he sent Morin on a journey of discovery to the mines of Hungary and Transylvania, a trip from which resulted Morin's first book, *Nova mundi sublunaris anatomia,* an account of the interior of the earth, published in 1619 and dedicated to his former patron du Vair. While in the mines, Morin had noticed the unusual heat, and he wrote the book to offer an astrological explanation of it, referring to the influence of the stars.

Morin even then had astrological inclinations, to be sure. But upon his return to Paris, they were strengthened. Morin made the astrological prediction that Dormy was in danger of death or imprisonment. Sure enough, Dormy was carted off in 1617, though given Dormy's involve-

---

1. The biographical sketch that follows is largely taken from an anonymous biography that appeared not long after Morin's death, *La vie de Maistre Jean Baptiste Morin* (1660). The editors of the Mersenne correspondence identify the author as Guillaume Tronson; see Mersenne 1932–88, 3:127–28. I also made use of the excellent biblio-biographical sketch, Martinet 1986.

ment in court politics, one would not have to have been a master astrologer to have made that prediction. But Morin was further confirmed in his vocation, and he went on to make numerous celebrated predictions, some of which were actually borne out.[2] After Dormy, Morin passed first to the patronship of the abbé de la Bretonnière, with whom he spent four relatively quiet years. In 1621 he passed on to the service of the duc de Luxembourg, brother of the favorite of the king, the duc de Luynes, who soon fell from favor. During this period Morin composed a number of works, including two astrological tracts and, in 1624, an interesting pamphlet attacking a group of young scholars who had announced a public disputation in which they proposed to refute the foundations of Aristotle's natural philosophy and replace it with a form of atomism.[3] In the latter tract Morin came out in favor of form and matter and against atoms—indeed, against innovation in natural philosophy in general. In the astrological tracts he came out against Copernicus and in favor of Tycho, though in general he showed his interest in bringing astrology up to date by making it consistent with the latest discoveries in observational astronomy, in particular the discovery and mapping of the southern skies.[4]

Though Morin believed in the guidance of the stars, he did not leave himself to their care alone. From all evidence, he was a firm believer in freedom of the will and the value of self-promotion. And so, when a chair of mathematics came open in 1629, he made himself a candidate, and with the influence of the cardinal de Bérulle and the Queen Mother, he received the appointment, which he held until his death in 1656. (The documents I have read imply that it was not his mathematical talent alone that won him the chair.[5]) It was during this period that Morin wrote most of his voluminous writings. Two extended disputes stand out. The first concerned a scheme that Morin proposed as early as 1633 for determining the longitudes of vessels at sea, a discovery that Morin hoped would

2. See *Vie de . . . Morin* (1660, 62–91) for an account of Morin's predictions. See also the entertaining, though not altogether reliable, account in Soprani 1987, 175–82. Morin's most disastrous prophecy was the incorrect prediction of the imminent death of Gassendi in the course of a pamphlet war.

3. The two astrological works are *Astrologicarum domorum cabala detecta* (1623) and *Ad australes et boreales astrologos; Pro astrologia restituenda epistolae* (1628). The polemical work is *Réfutation des thèses erronées d'Anthoine Villon . . . & Estienne de Claues . . .* (1624).

4. See Knappich 1986, 229–33.

5. See Bérulle's letter to Richelieu, in Mersenne 1932–88, 3:501–2. On the objections to Morin's elevation to the Collège Royal, see, e.g., Sédillot 1869, 101. Abuses of this sort seem to have been quite common at the Collège Royal in the early seventeenth century; see Goujet 1758, 1:206ff.

win him a pension from Cardinal Richelieu. Though the method worked in theory, it turned out to be not altogether practical, and despite roughly fifteen years of pleas and pamphlets, Morin never got his pension. The second large controversy was with Pierre Gassendi, over Morin's critique of Gassendi's atomism and Gassendi's critique of Morin's astrology, a dispute that began in earnest in the late 1640s. Also important from these years are two volumes attacking Copernican astronomy and the *Quod Deus sit* of 1635, to which we shall return. Through these years, though, starting in the early 1630s and extending up to the time of his death, Morin was working on his magnum opus, the *Astrologia gallica*, a *Summa astrologica*, as it were, a work that summarized his career as a natural philosopher and astrologist, published at The Hague in 1661, five years after his death. Though it was directed mainly at outlining his astrological system, this thick tome begins with a proof for the existence of God (a later version of his *Quod Deus sit*) followed by a series of books on natural philosophy, laying the foundations for the more properly astrological questions to follow. The natural philosophy that Morin outlines there is definitely conservative. He explicitly attacks mechanist and atomist views like those of Descartes and Gassendi. While there are some modern elements—a theory of space that looks as though it is derived from Patrizi, for example—Morin grounds his physics in the theory of substantial forms.

But yet in a sense Morin considered himself a sort of progressive. When Descartes' *Discourse and Essays* came out in 1637, Morin was one of the savants who received a copy. Descartes had hoped to collect a variety of responses to his first publication and publish them together with his responses, much as he was later to do with the *Meditations*. The whole exchange is quite interesting, and I have discussed it elsewhere.[6] For the moment, I would like only to note something that Morin said in his first reply to Descartes. In his letter of February 22, 1638, Morin wrote: "I do not know, however, what I should expect from you, for I have been led to believe that should I discuss matters with you using the terms of the Schools ever so little, you would immediately judge me more worthy of scorn than of response. But in reading your discourses, I find that you are not as much of an enemy of the Schools as you are made out to be." Morin continues with some remarks about his own view of the Schools: "The Schools appear to me to have erred only insofar as they are more occupied in speculation directed toward the search for the terms that we must use to discuss things, than they are in the search for the truth itself

6. See Garber 1988.

about things through good experiments; so they are poor in the latter and rich in the former. That is why I am like you in this matter; I seek the truth about things in nature alone, and I no longer put my trust in the Schools, which serve me only for terminology."[7]

This may strike us as something of a distortion, and it is, in a sense. But one can also see what Morin means. I mentioned Morin's trip of discovery to Hungary and Transylvania to visit the mines. In the preface to his *Anatomia* of 1619 Morin discusses the motivation for his explorations. The great diversity of opinion among the learned forms more of an obstacle to learning than a help. And so, he argues, we must turn away from books and to nature itself to discover how things really are. In doing so, Morin thinks that he has found an account of the makeup of the earth that is utterly unknown among the philosophers of the Schools.[8] I am sure that Morin saw this as exactly parallel to Descartes' rejection of learning and his travels in search of experience. Like Descartes, Morin professes to be following reason, not authority. In the *Astrologia gallica* he writes: "In what is said below we shall not follow the doctrines of the Schools, which are often in error, but we shall look to the nature of things, which alone contains the truth."[9] Even his treatment of astrology shows his open-mindedness. Though he agrees with the tradition that the stars influence what happens here below, he is not dogmatic about the details and sees the need to revise traditional astrological doctrines in the face of newly discovered astronomical facts. His is a progressive astrology, so to speak.[10]

At the same time, Morin's instincts are undeniably conservative. In doctrine, he follows Aristotle and opposes atomism and Copernicanism; at root, the traditional philosophy is right, if not in every detail. He is conservative in other respects too. A social climber of sorts, always looking out for a way to advance himself socially and financially, he vigorously opposes challenges to the institutions whose support and patronage he constantly sought. This, I think, is at least in part behind the vigor of the attack he made in his relative youth against a motley crew of anti-

7. AT 1:541.

8. See Morin 1619, dedication (unpaginated), letter to the reader, and esp. chap. 5. Morin opens the letter to the reader with a frank declaration of the novelty of his view: "Hic habes . . . Novam Mundi sublunaris divisionem, novas divisionis causas, novaque de rebus physicis disserendi fundamenta."

9. "Neque in infra dicendis sequemur doctrinas scholarum quae frequentius fallunt, sed naturam rerum spectabimus, quae sola veritatem continet" (Morin 1661, 39).

10. This is the main project of *Ad australes et boreales astrologos* of 1628. Of particular concern to him are the recent observations of the southern sky and how they affect astrology.

Aristotelians in 1624, in support of the government's condemnation and exile of three young scholars who proposed publicly to refute Aristotle, along with Paracelsus and the Cabala.[11] Though he considers himself open-minded, he has clearly hitched his star to the traditional philosophy of the Schools.

Morin, Mersenne, and Descartes

In general I find Morin to be a fascinating character. While his instincts are conservative, he is not a dogmatic Schoolman (I wonder whether anyone really was), and while he is a sort of progressive, he is no Descartes or Gassendi. But interesting as it would be to continue this discussion of Morin's life and works, we must turn now more specifically to his relations with Descartes and with Mersenne, Descartes' friend and the collector of the Second and Sixth Objections.

By the time Descartes had finished the *Meditations* and began to circulate it for comments, Descartes, Morin, and Mersenne had known one another for quite some time. Mersenne, too, had opposed the anti-Aristotelians of 1624 in print, as Morin had, and no doubt they became acquainted then, if they did not know one another before.[12] Descartes' acquaintance with Morin is usually dated from 1626 or 1628, and Morin is known to have helped Descartes get a piece of optical equipment made in the late 1620s.[13] Though Mersenne always opposed astrology and came to support Copernicanism, he always seemed to consider Morin a friend. When Morin published his arguments against Copernicus, Mersenne, along with Gassendi, counseled him against publishing the book, but after it was published, Mersenne did not disown him.[14] He even sent

11. In the 1624 pamphlet *Réfutation,* he complains more than once of the arrogance of the attack on Aristotelian thinking in the great city of Paris: "Ils [i.e., Villon and de Clave] afficent . . . un defi publique à toutes les Escoles, sects & grands Esprits . . . Et cecy non dans un village, mais dans une ville de Paris, à la face de la Sorbonne, de toute l'université, & du plus fameux Senat qui soit au monde" (6). Morin goes on to say that one of the reasons why he is attacking Villon and de Clave publicly is "pour l'honneur de ceste Cité tres celebre de Paris" (19–20).

12. See Mersenne 1625, 76–84, 96–113.

13. See Adam 1910, 90; and Mersenne 1932–88, 2:420. References to Morin appear in letters of 1629 and 1630: see AT 1:33, 124, 129–31, etc. In a letter dated February 22, 1638, Morin begins by recalling his earlier acquaintance with Descartes in Paris, presumably before Descartes left for Holland in 1628; see AT 1:537.

14. See Gassendi to Joseph Gaultier, July 9, 1631, in Mersenne 1932–88, 3:173.

the book to Descartes, whose reaction was, predictably, caustic.[15] Still, Descartes and Mersenne sought Morin's opinion of Descartes' *Discourse and Essays* when the book came out in 1637. In this interesting exchange, Mersenne acted as a sort of intermediary. He was with Morin when he received Descartes' first letter, in reply to Morin's, and he continued to encourage Morin to think of Descartes as someone knowledgeable in the School philosophy and not unsympathetic to it, and to encourage Descartes to continue to give Morin this (false) impression.[16]

Morin was clearly a friend of Mersenne's. He may not have been a member of the circle who met regularly in Father Mersenne's room in the Minim Couvent near the Place Royale.[17] He is, though, certainly a reasonable candidate for membership in the anonymous group of people who contributed to the Second and Sixth Objections, which Mersenne collected.[18] But one can go further than that. It can be established with reasonable certainty that Morin was a part of that group of objectors, and something plausible can be said about which specific objections he might have contributed to the enterprise.

Most relevant here is Morin's short treatise *Quod Deus sit* of 1635. Briefly (we shall look into it more carefully below), Morin's book is an argument in geometrical style for the existence of God.[19] Starting with a series of formal definitions and axioms, the book comprises thirty theorems purporting to establish the existence of God and his relation to the world. Mersenne knew this book and seems to have thought well enough of it to call it to Descartes' attention. He sent Descartes a note, now lost, apparently summarizing one of the arguments of Morin's *Quod Deus sit;* Descartes responded on November 11, 1640.[20] Descartes' response to

15. Descartes to Mersenne, summer of 1632[?], in AT 1:258.

16. See Garber 1988.

17. So argues Bernard Rochot. See his comment in Mersenne 1932–88, 10:410n. I'm not sure that Rochot is right.

18. There may be some precedent for Morin and Mersenne collaborating on a critique of Descartes. Baillet reports that some of the objections to the *Discourse* and *Essays* that Morin sent Descartes may actually be due to Mersenne: "Le Père Mersenne sembloit avoir joint quelques-unes de ses difficultez avec les objections de M. Morin" (1691, 1:356).

19. See Iwanicki 1936. Iwanicki offers a good discussion of the texts, Morin's arguments, Descartes' critique of the arguments, and the history of geometrical arguments for the existence of God. He also notes the connection between Morin and the Second Objections. However, Iwanicki sees himself as rather an advocate for the historical and philosophical importance of Morin, and he winds up greatly exaggerating Morin's place. Indeed, he concludes rather implausibly that Morin, not Descartes, has the best of the exchange on proofs for the existence of God.

20. AT 3:233–34.

Mersenne contains a critique of an argument that does not correspond exactly to anything that I can find in Morin's book itself, and it is impossible to evaluate Descartes' criticism without seeing exactly how Mersenne represented the argument. But the book itself followed shortly thereafter; it came from Mersenne via Huygens and arrived on January 21, 1641.[21]

Later we shall look more carefully at Descartes' response. But for the moment I would like to note only that Morin's name, and the name of his pamphlet, almost certainly appeared in the first version of the Second Objections that Descartes received. Writing to Huygens on January 16, 1641, Descartes notes: "I have been very eager to see the book, *Quod sit Deus,* because it is cited in the objections that Father Mersenne wrote you that he would send me."[22] The point at which it may have been cited is relatively easy to determine. At the end of the Second Objections we find the following passage: "After giving your solutions to these difficulties it would be worthwhile if you set out the entire argument in geometrical fashion, starting from a number of definitions, postulates and axioms. You are highly experienced in employing this method, and it would enable you to fill the mind of each reader so that he could see everything as it were at a single glance, and be permeated with awareness of the divine power."[23] While this passage does not make direct reference to Morin and his book (no passage in the final published text does), it seems quite plausible that this is the passage to which Descartes refers in his letter to Huygens. It is not altogether clear why Morin's name was dropped from the final published text. It could be that Descartes made it a general policy not to mention living authors by name. Another factor may have been the fact that Descartes was not impressed with the book. Rather than saying something uncomplimentary about Morin, thus offending him, something that he explicitly told Mersenne he did not want to do, Descartes may have decided to drop the reference altogether.[24] Perhaps, too, he did not want to start the sort of pamphlet war in which the somewhat irascible Morin had been known to engage with relish. Be that as it may, it seems reasonably certain that Morin and his *Quod Deus sit* stand behind this passage of the Second Objections.

With a little imagination, we may also be able to see Morin's hand in other passages of the *Objections* that Mersenne is known to have assembled. Morin's *Astrologia gallica,* a vast, encyclopedic work, begins with a full account of metaphysics and physics, in order to ground the

---

21. AT 3:283.   22. AT 3:765–66.   23. AT 7:128; CSM 2:92.
24. See Descartes to Mersenne, Jan. 28, 1641, in AT 3:294.

astrology in the later sections. Scattered throughout these introductory sections are passages criticizing Descartes, both his metaphysics and his physics. While Descartes' name occurs often in these pages, especially interesting is a critique that appears at the very beginning of the book, in a section entitled "Liber primus: De Vera cognitione Dei ex lumine Naturae; per Theoremata adversus Ethnicos & Atheos Mathematico more demonstrata." This, not unsurprisingly, is an expanded version of the pamphlet of 1635, the *Quod Deus sit*. (It had been published in 1655 as a separate pamphlet.) As we shall see, a number of interesting changes were made to the 1635 version, perhaps as a response to Descartes' criticism of that earlier work.

The most significant change is in the preface. While much of the preface of the 1635 version is retained, Morin added much new material, in fact a long and rather critical discussion of Descartes' metaphysics and his proofs for the existence of God. What is interesting here is a certain correspondence between the criticisms in this new preface and the contents of the Second Objections. Morin refers to the Second Objections and Descartes' replies a number of times, and to other sets of objections only rarely. In *Astrologia gallica* the first serious objection to the *Meditations* is the objection that the certainty of the *cogito* presupposes the proof for the existence of God. This, in essence a version of the circle objection that Arnauld brings up in the Fourth Objections in a somewhat more direct way, is the third point in the Second Objections.[25] Morin objects to the innateness of the idea of God and Descartes' causal principle, which corresponds to the second point in the Second Objections.[26] Morin also challenges Descartes' version of the ontological argument, a challenge that appears as the sixth point in the Second Objections.[27] Now, Morin does not bring up other points from the Second Objections, and he does bring up a number of other criticisms that are not found anywhere in the *Objections,* either in the Second Objections or in any others. But the correspondence between Morin's later criticisms in the *Astrologia gallica* and specific sections of the Second Objections suggests to me that Morin may well have contributed those objections to the pool, and that even if Mersenne wrote the final text, Morin may have formulated the criticisms. (It is interesting that there is no correspondence at all between Morin's critique in the *Astrologia gallica* and the Sixth Objections. This suggests to me that Morin probably had no hand in the later set.) There is a certain amount of conjecture in my claim that Morin may have been responsible

25. Morin 1661, 2; AT 7:124–25.   27. Morin 1661, 5; AT 7:127.
26. Morin 1661, 4; AT 7:123–24.

for these parts of the Second Objections, and I do not want to insist on it. Morin was self-important enough that I suspect he might have told his readers that he was behind those parts of the Second Objections if indeed he was. In the *Astrologia gallica,* for example, he certainly claims credit for having elicited the geometrical appendix to the Second Replies that is the answer to the request at the end of the Second Objections for a geometrical development of the main arguments.[28] So while I think that it is quite possible that Morin did contribute these objections, it is by no means certain.

Descartes' Critique of Morin

I have set out the facts, so far as I can establish them. Morin stands behind the Second Objections, in a sense; his *Quod Deus sit* was known to Descartes, and Morin is probably responsible for the request that Descartes present his arguments in geometrical form. Furthermore, it is possible (though not certain) that Morin made other contributions to the Second Objections, which Mersenne assembled. These are the facts. But what about the philosophy? While a great deal could be said here, I would like to concentrate on the issues raised by the geometrical proof for the existence of God that Morin offers in the *Quod Deus sit,* by Descartes' response to this work, both in his letters and in the Second Replies, and by Morin's response to Descartes' response to his *Quod Deus sit* in his *Astrologia gallica.*

From the introduction to Morin's booklet, his letter of dedication to the Sacra Comitia Gallica, the assembled clergy of France, who gathered in Paris in 1635, one can see why Mersenne would have found Morin's project sympathetic.[29] In that introduction, Morin explicitly opposes himself to "the unchecked sect of the atheists, who now have become so haughty."[30] He does so by appealing to the certainty of mathematical method to prove to atheists that they are constrained to accept the existence of God. He writes:

28. See Morin 1661, 5. See also Morin's *Defensio . . . dissertationis,* where he directly implies that he was behind Mersenne's question to Descartes: "Cartesius fuerit provocatus a . . . R. P. Mersenno, ut simili methodo conaretur demonstrare existentiam Dei, . . . nec tamen viris doctis satisfaceret" (Morin 1651, 90).

29. A dedication to the same group also appears in the second edition of 1655, on the occasion of the next meeting of the group in Paris. Bayle suggests that Morin had hoped to gain a pension from that group; see Bayle 1720, 3:2015 (s.v. "Morin," n. H).

30. Morin 1635, 4.

I have never doubted that one could show, not *what* God is, but *that* he is by way of the most evident light of nature alone. I also grasped that the greatest good deriving from this lies in the fact that the natural light still remains to the atheists, though they resist it, and with the help of that alone, they remain capable of grasping the first principles of nature, which they cannot fail to perceive even while they are denying them, because they are the *per se* objects of that [natural] light. Consequently, at the very least this path for discussing the existence of God with atheists is open to us, so that they might know their greatest error. Therefore, having undertaken this task for the glory of God, for the confirmation of faith, and to return the atheists to their senses, using a mathematical method, I carried it out to such an extent that once they concede those things I laid down as principles, perceptible by the light of nature alone, atheists cannot deny that God exists, that he created this world in time, and that he governs it by his providence, unless they themselves also deny that they exist.[31]

(Morin, like Descartes, appeals frequently to the light of nature.) Morin's introduction resembles Mersenne's project in the commentary on Genesis from 1623, which begins with a ferocious attack on the atheists of his day, followed by a multitude of arguments for the existence of God drawn from every conceivable premise.[32] It also resembles the letter of dedication to the doctors of the Sorbonne, which precedes Descartes' *Meditations,* emphasizing the necessity of refuting atheism by proving God's existence.[33] Like Descartes, Morin argues that knowledge of the existence of God is foundational for all other knowledge. He claims that many other important theorems can be derived from the ones he gives. Indeed, Morin claims: "It is not difficult to extend the principles I posit, and the theorems I set out to many other wonderful theorems about God and his creatures; indeed, . . . using the same method, one can prove the immortality of the soul and all the natural sciences."[34] In this respect Morin seems to resemble Descartes and the project of the *Meditations.* One can see why Mersenne would have thought that Descartes would find this sympathetic. Mersenne, in sending him the booklet, no doubt thought that he would draw Morin and Descartes closer together; it is interesting (and

31. Ibid., 5–6.
32. Mersenne 1623. The proofs for the existence of God begin on col. 25, and the subject is not set aside until cols. 669–74, where Mersenne presents a long diatribe against atheism. In all, Mersenne offers thirty-five arguments for the existence of God.
33. AT 7:1–2.
34. Morin 1635, 7.

perhaps revealing of Mersenne's character) that he miscalculated so badly. Despite their superficial similarity, Descartes found Morin to be quite a different kettle of fish.

Morin's *Quod Deus sit* contains a total of thirty theorems, but the argument for the existence of God is really quite simple and can be found in Theorems 14–16. Theorem 14 reads: "Omne ens finitum habet esse ab Ente infinito", "Every finite thing has its being from an infinite thing." [35] Morin offers two proofs for this theorem. The first is a direct regress argument. "Whatever there is must derive either from itself *[sit seipso]* or have its being from something else." This is one of Morin's axioms. But as a finite thing, something cannot derive from itself ("nullum ens finitum est seipso"); this is true by Morin's Theorem 12. Since there cannot be an infinite regress of causes or a circle (Prop. 13), there must be, somewhere in the series, an infinite cause. Thus every finite thing must have its being from an infinite cause, either directly or mediated by other finite causes. The second proof is somewhat different. Morin begins with a curious proof that would seem to establish that there can only be a finite number of finite things. Consider the number of people. Suppose that it is infinite. Then, Morin argues, it will contain all people who were, are, and will be, by his Axiom 10 ("There can be nothing greater or larger than the infinite, nor can any such thing be conceived" [36]). But, Morin notes, experience shows that new people are born every day. This, he infers, could not happen if the number of people were infinite, since, presumably, one cannot add anything to a number that is already infinite. Thus, the number of people must be finite. [37] But since each finite thing needs a cause, it follows that there must be something that is not finite that is the cause of everything else. And so, again, every finite thing has its being from an infinite cause.

Theorem 15 then establishes that "mundus hic finitus est," "this world is finite." The principal argument for this conclusion is grounded on Morin's refutation of the Copernican claim and his view that the earth is at rest in the center of the universe. Now assume that the universe is infinite. If so, the universe would occupy an infinite space. Since the universe turns around the earth once every twenty-four hours, it would then follow that matter would traverse an infinite space in a finite amount of time, which is absurd. And so, Morin concludes, the universe must be finite.

From these two theorems, it follows directly that there must be an infi-

35. Ibid., 15.
36. Ibid., 10.
37. Ibid., 16–17. A similar argument is found later, in Theorem 27.

nite being. For if finite things have their being from something infinite (Theorem 14) and if this universe is finite (Theorem 15), then there must be an infinite being. This is Morin's Theorem 16.

The earlier theorems deal with more general questions about infinite and finite beings. Before establishing that God exists, Morin establishes, for example, that the infinite being is *purus actus* (Theorem 2), that there are not two infinite beings (Theorem 5—shades of Spinoza), that the infinite being is indivisible and simple (Theorems 6 and 7), and so on. After establishing that the infinite being exists, Morin establishes that the infinite being produced everything by a simple act of will (Theorem 21), that the infinite being continually produces and conserves the finite beings he produces (Theorem 22), that the world was created in time (Theorem 27), and, finally, that the infinite being is the ultimate end *(finis)* of all finite beings (Theorem 30).

Needless to say, Descartes was not altogether impressed with this. He didn't expect much to start with. His dealings with Morin on the subject of his *Dioptrics* mostly left him unimpressed. Writing to Mersenne on December 31, 1640, Descartes noted: "I would not be unhappy to see what M. Morin has written about God because you say that he proceeds as a mathematician. But just between you and me, I don't expect very much of it, since I have never before heard of him involving himself with a writing of this kind." [38] Descartes' expectations were not disappointed. When he finally received the book, shortly after writing this note to Mersenne, he must have read it immediately. In his letter of January 28, 1641, Descartes transmitted his comments on the book to Mersenne:

> I perused M. Morin's little book. Its main shortcoming is that throughout he treats infinity as if his mind were above it, and could comprehend its properties. This is a shortcoming common to almost everyone, which I have carefully tried to avoid, since I have never treated infinity except to submit myself to it, and not to determine what it is or what it is not. Then, before giving any explanation of controversial matters, in his sixteenth theorem, where he begins to try to prove that God exists, he bases his reasoning on his purported refutation of the motion of the earth, and on the claim that the entire heavens move around it, something that he hasn't proved at all. And he also assumes that one cannot have an infinite number there, etc., something that he doesn't know how to prove either. Thus, everything he sets out right up until the end is quite far from being evident and quite far from the geometrical certainty that he seems to promise at

38. AT 3:275.

the beginning. I say this just between ourselves, if you please, because I don't want to displease him at all.[39]

Descartes' reaction here seems quite fair. Morin's proofs, like those of Spinoza, who would later offer a very different geometrical proof for the existence of God, are strongly based on the notion of infinity; Morin's God is from the first and primarily an infinite being, and it is on this divine attribute that Morin's arguments are grounded. And Descartes is certainly correct to note that Morin makes some very odd statements about infinity. Furthermore, Descartes correctly observes that Morin's geometrical proof in the *Quod Deus sit* depends crucially on the nongeometrical premise that the earth is at rest in the center of the universe.

This was the last time Descartes mentioned Morin in his correspondence, at least in that which survives, and it is the only passage in which Descartes explicitly addressed Morin's book. But it seems to me that much of what Descartes has to say about the geometrical mode of exposition in the Second Replies is also directed specifically against Morin. Descartes begins by distinguishing between two things, the order of exposition, the *ordo scribendi,* and the way of demonstrating, the *ratio demonstrandi.* To write in order is simply to write in such a way that "the items which are put forward first must be known entirely without the aid of what comes later; and the remaining items must be arranged in such a way that their demonstration depends solely on what has gone before." The *ratio demonstrandi,* on the other hand, is twofold and represents two different ways of realizing order. The *ratio* of analysis "shows the true way by means of which the thing in question was discovered." This, Descartes tells us, is what he used in the *Meditations.* "Synthesis, by contrast . . . demonstrates the conclusion clearly and employs a long series of definitions, postulates, axioms, theorems and problems, so that if anyone denies one of the conclusions it can be shown at once that it is contained in what has gone before."[40] By *synthesis* here Descartes clearly means quite specifically the sort of method that Morin used in *Quod Deus sit,* a quasi-geometrical demonstration using definitions, axioms, and so on.

Descartes makes no bones about it: analysis is vastly to be preferred to synthesis, at least in metaphysics. He writes:

> In metaphysics . . . there is nothing which causes so much effort as making our perception of the primary notions clear and distinct. Admittedly, they are by their nature as evident as, or even more evident than, the primary notions which the geometers

39. AT 3:293–94.
40. AT 7:155–56; CSM 2:110–11.

study; but they conflict with many preconceived opinions derived from the senses. . . . And so only those who really concentrate and meditate and withdraw their minds from corporeal things, so far as is possible, will achieve perfect knowledge of them. Indeed, if they were put forward in isolation they could easily be denied by those who like to contradict just for the sake of it. This is why I wrote "Meditations" rather than "Disputations," as the philosophers have done, or "Theorems and Problems," as the geometers have done.[41]

This last phrase seems to be a clear reference to Morin's pamphlet. And a few lines later there is another, I think: "I am therefore right to require particularly careful attention from my readers; and the style of writing that I selected was one which I thought would be most capable of generating such attention. I am convinced that my readers will derive more benefit from this than they will themselves realize; for when the synthetic method of writing is used, people generally think that they have learned more than is in fact the case."[42] Descartes thus has very little regard for the use of the geometrical or synthetic style of writing in metaphysics. At best, he argues, it is a style appropriate for geometry, where "the primary notions which are presupposed for the demonstrations of geometrical truths are readily accepted by anyone, since they accord with the use of our senses." But, it should be noted, Descartes is not even particularly happy with the use of the geometrical style of writing in geometry. He writes: "It was synthesis alone that the ancient geometers usually employed in their writings. But in my view this was not because they were utterly ignorant of analysis, but because they had such a high regard for it that they kept it to themselves like a sacred mystery."[43]

Synthesis thus seems to be good for very little. (This is yet another reason to be suspicious of the often-made claim that Descartes voluntarily decided that he was going to write his *Principles* in the synthetic style.) But yet, Descartes goes ahead and responds to the request of the authors of the Second Objections and presents his arguments in the style of the geometers. Given what he said about synthesis, this is not a little puzzling. Granting the difficulty of his *Meditations,* Descartes tells his readers that he is giving them this morsel of the argument not as a substitute for the analytic *Meditations,* but in order to give them help in comprehending some specific arguments that are particularly difficult and particularly important.

41. AT 7:157; CSM 2:111–12. 43. AT 7:156; CSM 2:111.
42. AT 7:158–59; CSM 2:112.

But even this example of synthesis is an implicit critique of Morin's procedure in *Quod Deus sit*. Like Morin's book, Descartes' geometrical arguments have definitions, axioms, and theorems. In general it is not illuminating to compare in detail Descartes' text with Morin's. Unlike Morin, Descartes seems to do his best to avoid the notion of infinity.[44] The propositions simply formalize arguments found already in the *Meditations;* there seems little there that can be regarded as a specific reply to Morin's pamphlet. But Descartes' geometrical exposition has something that Morin does not: postulates. In a standard Euclidean geometry there is little to distinguish postulates from axioms; in both cases we are dealing with propositions that must be assumed to do proofs. But in Descartes' geometrical arguments, the postulates are something else, not propositions at all:

> The first request I make of my readers is that they should realize how feeble are the reasons that have led them to trust their senses up till now. . . . I ask them to reflect long and often on this point. . . . Secondly I ask them to reflect on their own mind and all its attributes. . . . Fifthly I ask my readers to spend a great deal of time and effort on contemplating the nature of the supremely perfect being. Above all they should reflect on the fact that the ideas of all other natures contain possible existence, whereas the idea of God contains not only possible but wholly necessary existence. This alone, without a formal argument, will make them realize that God exists.[45]

These are hardly postulates of the usual sort. They are in fact demands, as the Latin *postulare* would suggest, things we are asked to do, not merely to accept. In including such postulates in his geometrical presentation, Descartes is answering the criticisms of the geometrical mode of writing he made in the Second Replies; it is only because he includes such postulates, Descartes thinks, that the geometrical mode of presentation is capable of leading us to knowledge of things metaphysical. In this way, the differences between Descartes' and Morin's geometrical arguments for the existence of God simply underscore Descartes' rejection of Morin's chosen form of presentation. Thus the geometrical presentation that follows the Second Replies can be read not only as a clarification of the arguments, terminology, and assumptions used in the *Meditations,* not only as a civil answer to a civil question from the authors of the Second

---

44. However, note Descartes' Axiom 6, in AT 7:165–66.
45. AT 7:162–63; CSM 2:114–15.

Objections, but also as a philosophical exercise directed against the *Quod Deus sit* of Jean-Baptiste Morin.

Morin's Response

The Second Replies is the last text in which Descartes has anything to say about Morin; as far as Descartes was concerned, the less said, the better. But though Descartes may not have had anything more to say about Morin, Morin had quite a lot to say about his more famous colleague.

The response is found in Morin's posthumously published *Astrologia gallica*. While there is no direct evidence that Mersenne actually showed Morin the direct criticisms Descartes made of his work, the paragraph in the letter quoted above, the alterations Morin made in the new edition of the *Quod Deus sit*—included in the *Astrologia*—suggest that Mersenne may well have transmitted the essence of those criticisms. Though in the end he does not give up his strong dependence on infinity, nor does he actually alter many of the details of his proofs, the rearrangements and the additional axioms and definitions show some sensitivity to Descartes' concerns.[46] Also, later in the *Astrologia* there is considerably more discussion of Descartes, particularly his physics. Altogether, this amounts to an additional set of objections against the *Meditations,* and against the *Principles* too, one especially worth study given Morin's rather interesting position in the intellectual world of mid-seventeenth-century France. But rather than trying to survey the whole of Morin's attack against Descartes, let me just touch on a few issues with respect to the questions of analysis versus synthesis and Descartes' geometrical arguments.

Morin begins his discussion of Descartes' geometrical exposition by noting that it was he, Morin, and his *Quod Deus sit* that elicited the discussion:

> Although my little book against the atheists pleased everyone, after the publication of his *Meditations,* those who were not satisfied with his demonstrations for the existence of God through our idea of him requested Descartes to prove the same a posteri-

46. For example, in the new version of Theorem 15, now Theorem 22 ("This world is finite"), Morin eliminates the argument he had used earlier, and that offended Descartes so much, which depends upon his refutation of Copernicus. See Morin 1661, 11. (This same offensive argument, though, appears later in a different context; see Morin 1661, 51.) Similarly, in the new version of Theorem 27, now Theorem 35, Morin eliminated the assumption about infinity that Descartes found so problematic, that one cannot add anything to an infinite number. See Morin 1661, 13. There are other, smaller changes as well that are suggestive.

ori through his creatures, as I had done. To that same end, that same little book was requested of me, which the Reverend Father Mersenne, known to all of the learned, sent him in Holland, so that he might see my method for proceeding in the geometrical fashion.[47]

Morin goes on to examine the three proofs that Descartes gives in his geometrical appendix, finding them, one by one, unsatisfactory. Most interesting, though, are his comments on Descartes' remarks on the analytic and synthetic modes of reasoning. He criticizes Descartes' use of both ways of proceeding.

Morin notes that Descartes does try to give a geometrical account, like Morin's own, using definitions, axioms, and theorems. But he also takes note of the fact that Descartes makes use of postulates: "Then there are also seven postulates, by which the mind binds itself. However, I have demanded [postulaverim] nothing. Rather, I have left the mind with its freedom of judgment."[48] It is interesting here that Morin does not seem to understand exactly why Descartes adds the postulates in the way he does, nor does Morin understand the rather radical difference between Descartes' postulates and those more commonly found in the tradition. All he says is that they seem to bind the intellect in a way that he does not want to. He does continue, however, with a rather uncharacteristically penetrating critique of Descartes' Postulate 5: "I ask my readers to spend a great deal of time and effort on contemplating the nature of the supremely perfect being. Above all they should reflect on the fact that the ideas of all other natures contain possible existence, whereas the idea of God contains not only possible but wholly necessary existence. This alone, without a formal argument, will make them realize that God exists."[49] Morin comments: "Once we have conceded this postulate, then no definitions, no axioms, nor any demonstrations are needed, either through analysis or through synthesis."[50] Morin's point is a good one: take this particular postulate seriously, and there is no need for argument at all.

Morin does not discuss Descartes' general remarks on the preferability of analysis over synthesis for metaphysics; the general theoretical position seems to escape him. But he does say why he thinks that analysis is not an appropriate way of proving the existence of God. Morin writes:

47. Morin 1661, 5.   48. Ibid.   49. AT 7:163; CSM 2:115.
50. Morin 1661, 6. Morin goes on to say that if we do not concede the postulate, then we cannot pass from the idea of God in the mind to his existence in reality, but that is a longer story.

And indeed it seems remarkable to me that M. Descartes chose the analytic method for proving the existence of God, which is utterly inappropriate for this purpose. Analysis is defined by Viète as *the assumption of that which is sought as if it were conceded, then through consequences* [passing] *to that which is generally conceded as true.* If it is generally conceded as true that he [Descartes] exists from the fact that he thinks and, indeed, that he has an idea of an infinitely perfect being, which he calls God, then that which we seek will be whether God exists. Now, from the definition of analysis we should assume that God exists, as if it were conceded, and from that concession, we should seek [to show] as a consequence that M. Descartes, or he who has the idea of an infinitely perfect being, that is, God, thinks and therefore exists. But in his analysis, he demonstrates nothing of the sort; indeed, nothing of the sort can be demonstrated. For God exists from eternity, but M. Descartes has not thought from eternity, and therefore did not exist, nor did he have the idea of God [from eternity]. Therefore it is obvious that analysis can do nothing toward proving the existence of God from the idea of God which he says that he has, considering that idea as the concept of a being of greatest perfection or of infinite nature, as he often does.[51]

The criticism is just, if we assume that Descartes had in mind Viète's conception of analysis here. While it would take us too far afield to demonstrate this, I think that Morin's criticism shows that he simply misunderstood what Descartes was up to in calling the *Meditations* analytic, just as he missed the deeper points behind Descartes' critique of the geometrical mode of writing in metaphysics. It is quite clear that Descartes' *Meditations* are not intended to be analytic in the sense in which Viète's mathematics is.[52]

But Morin kept insisting, stubbornly, on the fact that he was right on this issue, as on others, and Descartes was wrong. His final proof was, in his eyes, definitive: the doctors of the Sorbonne gave his *Quod Deus sit* the approbation that they denied Descartes' *Meditations.* Here is an argument from authority if ever there was one.[53]

In this essay I have argued that Jean-Baptiste Morin and his *Quod Deus sit* stand behind at least parts of the Second Objections, and that it was

51. Ibid., 7.

52. For one interpretation of what Descartes means when he calls the *Meditations* analytic, see Garber 1986.

53. Jean-Robert Armogathe has recently argued that, contrary to what Morin thought, Descartes actually did receive the approbation of the Sorbonne. See Armogathe 1994.

specifically to Morin and his little book that Descartes was responding at the end of the Second Replies and in the geometrical appendix. But how does this change our understanding of those passages? Perhaps not at all; interesting as that bit of historical information may be to those of us with an antiquarian bent, it may not have any real philosophical bearing. But then maybe it does.

I would like to end with a kind of conjecture, a stab at an argument that one might make on the basis of my historical argument. I think that what I have presented here strengthens the case for saying that however important it might be for earlier thinkers, however much it may be emphasized by later commentators, the doctrine of analysis and synthesis may not be a central tenet in Descartes' own thought, not a basic category in terms of which Descartes liked to think of his work and that of others. Rather, I suspect that it is a very specific response to a very specific proposal for how to do metaphysics, a proposal embodied in the example of Morin's *Quod Deus sit*. And, I think, it is a clear rejection of that way of doing metaphysics. Even though Descartes does develop his ideas in synthetic form in the geometrical appendix to the Second Replies, it must be emphasized that this is largely (only?) to show the inadequacy of that form and the problems inherent in an enterprise of the sort that Morin was attempting to undertake. This does not establish for certain that Descartes did not then generalize the notion of synthesis, or take it seriously in his own later works. But, I think, the argument should somewhat undermine whatever temptation we might have to see synthesis as a more general category and to try to include the *Principles* as synthetic, as many readers from Martial Gueroult to Edwin Curley and J. M. Beyssade have done.[54] In late 1640 and early 1641, when Descartes confronted the geometrical argument of Morin and penned his response, both his private response to Mersenne and his more public response in the Second Replies, and when he began drafting what was to become the *Principles of Philosophy*, he saw nothing to recommend a geometrical metaphysics of the sort that Morin was attempting to establish.

54. For a more systematic attack on the idea that the *Principles* should be understood as synthetic, see Garber and Cohen 1982.

·❧ 6 ❧·

HOBBES'S OBJECTIONS AND HOBBES'S SYSTEM

Hobbes's objections to the *Meditations* belong to a period when he was cut off from his country and his patrons and was living in relative poverty in Paris. He suffered a kind of isolation, but he also acquired the leisure and confidence to develop his ideas. He was fifty-two when he came to Paris in 1640, and for nearly all of his adult life he had served in the households of influential English aristocrats. He worked for more than one branch of the Cavendish family, starting out as tutor to the earls of Devonshire but later entering the employment of their cousins, the earl of Newcastle and his brother. The services that he provided were those of an intellectual. He functioned as teacher, scientific consultant, translator, political adviser, and as a propagandist in the Royalist cause, and it was in the course of his duties that he first met Marin Mersenne in 1630 and, ten years later, fled England for Paris, where he became a member of Mersenne's circle.

During the English Civil War and for some time after it, Paris was the home of the English monarch-in-exile. His court had numerous aristocratic hangers-on, including the Newcastle Cavendishes. Hobbes kept up his connections with this group, but he no longer lived in their shadow. With the success of his book *De cive,* published in a limited edition in Paris in 1642, he began to make his name as a philosopher, and he was consulted by local savants on a wide range of subjects, from mathematics and natural philosophy to morals and politics. Hobbes's plan for his own system of philosophical ideas was first sketched in print during his Paris days. The system was to take the form of a philosophical trilogy concerning body, man, and citizen. It was the third installment of this trilogy, *De cive,* that enjoyed a favorable reception in Paris in 1642, and though it was the only part to be published in the 1640s, it was not the only one Hobbes worked on. He wrote drafts of *De corpore,* the first book of the trilogy, and he completed some pieces on optical topics that were relevant to its sequel, *De homine.* So the time that Hobbes spent in Paris was crucial for the articulation of a large part of his system.

Some commentators have wondered whether France provided more than the geographical setting for Hobbes's philosophical development between 1640 and 1650. They have asked whether the intellectual milieu in general, or Descartes' philosophical project in particular, actually inspired Hobbes's philosophy.[1] There is some evidence that Hobbes made a study of Descartes' *Discourse and Essays* between 1637 and 1640 and that he wrote at length about these works in a letter to Mersenne that is now lost. It is even possible to interpret parts of Hobbes's first full-length political treatise, completed before his flight to Paris, as an attempt to improve on solutions to metaphysical problems outlined in the *Discourse*.[2] On this interpretation Hobbes began philosophical writing in earnest as an objector to Descartes' *Discourse and Essays,* and he had already gone some way toward constructing a metaphysical system to rival Descartes' when he composed his objections to the *Meditations*.[3]

In what follows I will examine, and eventually reject, this interpretation. It seems doubtful to me that Hobbes's system was intended to address problems Hobbes associated with Descartes or first encountered through Descartes, and so I do not believe that it was in the role of objector to Descartes that Hobbes began to make his own way in philosophy. Starting with Hobbes's objections to the *Meditations,* I suggest that they bypass Descartes' main line of argument and often misrepresent Descartes' conclusions. If they are intended to present an answer to Descartes' problems to rival Descartes' own answer, then they fail to do so, and fail conspicuously. Instead of addressing the problem of making physics skepticism-proof or even the problem of proving that God exists and that the soul is immortal, Hobbes's objections suggest how Descartes might have recast some of his claims in ways that would be acceptable to a materialist. Hobbes shows very little interest in the consequences of Descartes' proofs of God or in the use that Descartes makes of his conclusions about the soul.

What of Hobbes's objections to the *Discourse and Essays?* Do these suggest that his conception of what is to be done in philosophy is derived from his dissatisfaction with Descartes? Again, I argue, the answer is no, especially in regard to what he thought had to be done in metaphysics. More important to his philosophical development than the content of his objections to the *Discourse and Essays* or the *Meditations* is his status as an objector, which depended on his being accepted as an equal among the recognized philosophers of Paris in the 1640s. It is being a prominent

1. Brandt 1928.   3. Sepper 1988.
2. Tuck 1988. See also Tuck 1989.

and active member of a stable group of intellectuals over a considerable period that matters to the development of his system, I believe, not some special impact of Descartes' project on Hobbes's thought. Even the influence of Mersenne and Mersenne's circle should not be exaggerated. Mersenne undoubtedly offered encouragement to Hobbes and opportunities to make his ideas better known, but this falls short of setting Hobbes's agenda. A wide variety of views and preoccupations is represented in Mersenne's circle, after all, and Hobbes's correspond to those of Mersenne loosely at best. The same is true of Hobbes and another member of Mersenne's circle, Gassendi. Indeed, for all the evidence shows, the elements of his system, while certainly elaborated in Paris, could have been inspired in another, smaller, and less formal intellectual milieu centered on some of his English aristocratic patrons. It could have been as a member of a network of correspondents centered on Charles Cavendish in the early 1630s, or as a participant in scientific discussions that took place while on one or another grand tour of the Continent in the 1630s or before, that Hobbes first began to contemplate the systematic treatment of body, man, and citizen. Hobbes probably played the role of objector or critic in these discussions and as a correspondent, and so it may have been as an objector that he first stumbled on his vocation as system builder. What does not seem to be true is that Hobbes came upon this calling as an objector to Descartes.

Hobbes's Objections and Descartes' Immaterialism

I begin with Hobbes's objections to the *Meditations*. Apart from Bourdin's, Hobbes's were the only ones to appear in print interspersed with Descartes' replies. In a way, this is the treatment they invite, for Hobbes does not present a sustained line of thought that the interspersed replies interrupt. He presents a list of points, which the replies take up one by one. Some of Hobbes's points are connected, but it is not clear how they add up, and it is hard to see in them a critical view of the *Meditations* as a whole. There are sixteen objections, and all but one are comments on short passages quoted from Descartes' text. Not all of the passages have a clear bearing on the topics of the individual meditations from which they are taken. If they engage a single theme of Descartes' thought, it is his immaterialism, immaterialism about thought in general, about the self, about conceptions of God, and about mathematical natures.

An objection that Hobbes directs against the Second Meditation sets the tone. He accepts that from the fact that I am thinking it follows that I exist, but he wonders whether Descartes can properly extract, as a corollary, that the "I" is a mind, or an intelligence, or a thinking thing. For all the *cogito* shows, the "I" could be corporeal. And not only does the *cogito* leave open the possibility of the I being corporeal, Hobbes goes on, but the later wax argument actually *shows* that the I is corporeal:

> We cannot conceive of jumping without a jumper, or knowing without a knower, or of thinking without a thinker.
>
> It seems to follow from this that a thinking thing is something corporeal. For it seems that the subject of any act can be understood only in terms of something corporeal or in terms of matter, as the author himself shows later [in] his example of the wax: the wax, despite the changes in its colour, hardness, shape and other acts, is still understood to be the same thing, that is, the same matter is the subject of all of these changes.[4]

Descartes' reply concedes that acts need subjects, that it is a thing that is hard, changes shape, and so on, and also a thing that thinks, but he insists that the "thing" in this sense is neutral between the corporeal and the spiritual. He insists, too, that he is noncommittal about the nature of the thing that thinks in the Second Meditation. Far from assuming that the thing that thinks is spiritual or immaterial, Descartes says, "I left it quite undecided until the Sixth Meditation, where it is proved."[5]

Is Descartes putting a partisan interpretation on his text when he says that in the Second Meditation he left undecided the nature of the thinking thing? Does he not say that the "I" of the Second Meditation is a mind, intelligence, or reason, and does this not tilt the answer to the question of his nature decidedly toward the spiritual? The truth seems to be that he attributes mentality or intelligence to himself, these being the only things he finds belonging to him when under the discipline of the method of doubt, but that he does not identify this mentality or intelligence as essential, and indeed does not deny that he might have a bodily nature. For while he supposes under the discipline of doubt that there are certain things he is not—a structure of limbs, for example, or a thin vapor permeating the body—he acknowledges that he is only supposing, and that the suppositions come naturally only after he has resolved to take as nonexistent anything whose existence can be doubted.[6] None of this stops him from reminding himself how tentative are his refusals to identify himself with the limbs or the vapor. For he asks, "And yet may it not perhaps be the case that these very things [i.e., the structure of limbs, the thin vapor] which I am supposing to be nothing, because they are unknown to me,

4. AT 7:173; CSM 2:122.   6. See AT 7:27; CSM 2:18.
5. AT 7:175; CSM 2:123.

are in reality identical with the 'I' of which I am aware?" And he answers, "I do not know."[7]

Descartes, then, does leave open what he says he leaves open, and if Hobbes misunderstands that, it may be because he misunderstands the rules of the method of doubt. While implementing the method, Descartes does assume rather than prove that there is no body for the thinking thing to be or for the thought to inhere in. But this is not a case of begging the question, for the belief in the existence of bodies is reinstated in the Sixth Meditation, and with it the question of whether the subject is essentially immaterial or material.

Hobbes attacks Descartes' immaterialism from another direction when he tries to show that it is not needed to underpin the distinction between imagination and conception by the mind. In his fourth objection, Hobbes equates imagination in Descartes' sense with having an idea of a thing, and conception in Descartes' sense with reasoning to the conclusion that something exists. Descartes already agrees that imagination is a partly corporeal process resulting from action on the sense organs, but his text suggests that conception by the mind is an altogether different kind of operation. Hobbes puts forward a suggestion that allows the explanation of conception and imagination to be linked, without the postulation of immaterial things. He proposes that reasoning is a process in which labels attached to various things are concatenated into sentences according to conventions agreed on by humans. Conception would depend not only on the conventions for joining the names into sentences but also on the imaged features of things for which names were given. Or, as Hobbes puts it, "[R]easoning will depend on names, names will depend on the imagination, and imagination will depend (as I believe it does) merely on the motions of our bodily organs; and so the mind will be nothing more than motion occurring in various parts of an inorganic body."[8]

The compatibility of this proposal with mechanistic explanation appeals to Hobbes; but Descartes raises some powerful doubts about Hobbes's idea that names alone come into reasoning.[9] He takes it, contrary to Hobbes, that reasoning is a matter of linking together the significations of names and also, and surely correctly, that the significations of some names cannot be imaged. This is the clear implication of his remarks concerning chiliagons in the Sixth Meditation, to which Descartes alludes in replying to Hobbes.[10]

As in the case of his objection concerning the subject of thinking, in

7. Ibid.
8. AT 7:178; CSM 2:126.

9. AT 7:179; CSM 2:126.
10. AT 7:72; CSM 2:50.

which Hobbes seems to misread or overlook the constraints of the method of doubt, in his objection to Descartes on imagination and conception he seems to miss the point. Descartes is not trying to explain the workings of the faculties that result in science only to find he has to explain some of them on immaterialist principles. He is trying to show that science is possible, that real knowledge of the physical world is available, because not all of our faculties can coherently be held to be unreliable. Conception by the mind is a case in point. It cannot be held to be unreliable, because it is autonomous of unreliable sense perception. Hobbes does not see that it is the objectivity of conception rather than the process of conception with which Descartes is concerned. And doubting the objectivity of conception himself, Hobbes does not seek to reconstruct conception as reasoning that might be guaranteed to lead to true conclusions; he wants only to reconstruct it in ways that will not multiply entities beyond those required by mechanistic explanation.

Hobbes challenges Descartes' immaterialism not only when he offers his own model of the genesis of conception, or his own anti-innatist account of the formation of the idea of God (fifth, seventh, and tenth objections), or the astronomical idea of the sun (eighth objection). Immaterialism is also his target when he queries Descartes' commitment in the Fifth Meditation to geometrical natures that can exist independent of the mind even in the absence of triangles: "If the triangle does not exist anywhere, I do not understand how it has a nature. For what is nowhere is not anything, and so does not have a being or nature."[11]

Hobbes goes on to suggest how, without assuming the existence of natures in the abstract, but by assuming only the existence of ideas of triangles and names for triangles and their properties, it is still possible to have eternal geometrical truths that in turn are material for geometrical demonstrations. Here, as in so many other instances in his objections, Hobbes appears to think that what is at stake in the *Meditations* is the economical explanation of the existence of some of our ideas and reasoning, in this case ideas and reasoning in geometry, and not the question of how the ideas and reasoning can amount to science in the sense of knowledge that withstands the most radical doubt. Immaterialism promotes Descartes' overall aim of answering the skeptic; Hobbes misses this function while realizing that Descartes' immaterialism is important.

The major theme of Hobbes's objections is Descartes' immaterialism; the minor theme is Descartes' traditionalism and Scholasticism. From the very beginning of his objections, Hobbes chides the author of the *Medita-*

11. AT 7:193; CSM 2:135. See AT 7:64; CSM 2:44–45.

tions for disappointing readers' expectations of originality, and he complains of the presence of phrases and lines of reasoning whose obscurity does not sit well with Descartes' repeated insistence on the clarity and evidence of his demonstrations. Hobbes's first objection is that in the First Meditation, Descartes has done no more than recycle old philosophical lore about the unreliability of the senses. Hobbes does not deny that the senses on their own are unreliable, but he does not think this lesson has to be taught yet again. "I am sorry," Hobbes says, "that the author, who is so outstanding in the field of original speculations, should be publishing this ancient material." [12] Later, in his third objection, Hobbes worries about the basis for ascribing a mental nature to the thing that thinks: "If M. Descartes is suggesting that he who understands is the same as the understanding, we shall be going back to the scholastic way of talking: the understanding understands, the sight sees, the will wills, and by a very close analogy, the walking (or at least the faculty of walking) walks. All these expressions are obscure, improper, and quite unworthy of M. Descartes' usual clarity." [13]

This is not the only place in the *Objections* where Descartes is accused of retreating into Scholastic obscurity. Hobbes wonders what it can mean to claim, as Descartes does in the Third Meditation, that some substances are more real than others, and he asks whether it is really permissible in argument to invest a premise with evidence by appeal to the presence of a great light in the intellect, as Descartes does in the Fourth Meditation. [14]

These accusations of traditionalism and Scholasticism seem to be ill-founded, importantly so in the case of the First Meditation, where Hobbes seems entirely to miss the point that Descartes' skepticism is methodological. But I think that Hobbes was sincere in making them and that they are of a piece with a remark about Descartes that Hobbes's friend and biographer, John Aubrey, was to quote. According to Aubrey, Hobbes said that had Descartes "kept himself to Geometry he had been the best geometer in the world but that his head did not lye for philosophy." [15] This remark, which may otherwise seem a colossal misjudgment or a piece of spite or rancor reflecting the bad personal relations between Descartes and Hobbes from an early date, makes sense if one attributes to Hobbes and Descartes different conceptions of philosophy, and if one takes seriously the evidence already reviewed that Hobbes did not grasp Descartes' conception of the subject after reading the *Meditations*. For

12. AT 7:171; CSM 2:121.
13. AT 7:177; CSM 2:125.
14. See AT 7:185; CSM 2:130. Also AT 7:58–59; CSM 2:41.
15. Aubrey, 1975, 168.

Hobbes, philosophy or science consists of any reasoning that systematically increases knowledge or probable belief about causes and effects for human benefit. For Descartes, this conception would apply, and even then only approximately, to parts of philosophy outside first philosophy. Though it is possible to describe first philosophy in Descartes' sense as a kind of causal knowledge, the knowledge, namely, of how all effects depend on God, it is more accurate to say that it is the knowledge of how all genuine knowledge of effects depends on knowledge of God and the soul. For Hobbes, there is no such branch of philosophy, and indeed no such fact as that knowledge of effects depends on knowledge of God. There is first philosophy, but this consists of the explication of the most general concepts required to explain different effects as the result of different combinations of matter in motion. As an objector to the *Meditations,* Hobbes tends to focus on passages in which Descartes appears to depart unnecessarily from explanations involving matter in motion or explanations involving matter, motion, imagination, and names. Hobbes tends to focus, that is, on passages in which his conception of philosophy can be applied in a dispute with Descartes, not passages, such as those involved in the controversy about the Circle, in which the tensions of Descartes' own conception are revealed. Naturally Hobbes had more to comment upon in the *Discourse and Essays* than in the *Meditations,* for the essays are full of philosophy in more or less Hobbes's sense, and the dependence of Descartes' philosophy on a system about God and the soul, as opposed to a few pronouncements about God and the soul, is far from evident.

Hobbes and Scepticism

I now leave the *Meditations* to consider the impact of the *Discourse and Essays* on Hobbes's thought. In 1637 Sir Kenelm Digby sent to Hobbes in England a copy of Descartes' *Discourse and Essays.* The book elicited detailed comments from Hobbes in 1640 in the form of a fifty-six-page letter to Mersenne. The letter appears to be lost, but excerpts from it that are critical of the *Dioptrics* are quoted in correspondence between Mersenne and Descartes, and some of these excerpts survive. The excerpts, as well as a manuscript work on optics whose date is uncertain, show that Hobbes made a careful study of the first of Descartes' *Essays.*[16] Is there any evidence that the other essays, or that the *Discourse* itself,

16. See British Library, Harleian MS 6797. An edition by F. Alessio appeared in *Rivista Critica di Storia della Filosofia* 18 (1963):147–288.

influenced Hobbes before the appearance of the *Meditations*? It has recently been suggested that, by 1641, and partly under the influence of the *Discourse and Essays,* Hobbes had formulated a comprehensive general philosophy to rival Descartes' own, a philosophy that set out in particular to dispel Descartes' hyperbolical doubt. According to this interpretation, there are extensive traces of hostility to Descartes in Hobbes's earliest full-length political work, *The Elements of Law* of 1640. Passages from the *Elements,* when read in the light of an optical treatise of Hobbes that may belong to the end of the 1630s, are supposed to contain nothing less than an answer to Cartesian skepticism that improves on Descartes' own answer.[17]

The main lines of the answer are supposed to come from chapter 2, article 10, of *The Elements of Law.* Hobbes there argues that in seeing color, in hearing sound, and "in the conceptions that arise in the other senses," the sensible quality inheres in the sentient creature and not in the external object. He concludes that "whatsoever accidents or qualities our senses make us think there be in the world, they are not there, but are seemings and apparitions only. The things that really are in the world without us, are those motions by which these seemings are caused."[18] Indeed, unless moving bodies were acting upon us to produce the seemings and apparitions, routine experience of qualitative change would be impossible. The inconceivability of experienced change in the absence of external objects, stressed in an optical manuscript sometimes known as the "Tractatus opticus," amounts for at least one commentator to a sort of proof of the external world given the experience of qualitative change, a proof that, unlike Descartes' proof, does not rely on an argument for God's existence.[19]

Does Hobbes in this argument take seriously Descartes' hyperbolical doubt while refuting it more economically than Descartes' arguments in the *Meditations*? It is hard to see how. Under the assumptions in force until it is actually dispelled, the hyperbolical doubt of the *Meditations* simply does not allow one to trace experienced change to the action of an external material body: the operation of the demon has to be invoked. Indeed, one does not enter into the hyperbolical doubt at all unless one grants Descartes that all experience could be produced by the demon in the absence of moving bodies. Far from rescuing anyone from Descartes' doubt, the argument from *The Elements of Law* simply violates its ground rules.

17. Tuck 1988, 37ff.   19. Tuck 1988.
18. Hobbes 1969, 7.

It does so, at any rate, if it is *supposed* to rival Descartes' way of clearing up the hyperbolical doubt. But it is far from clear that it is meant to do so. First, the argument in the *Elements* is basically about how we can be mistaken in sense experience; in the *Discourse,* on the other hand, Descartes has just as much to say about mistakes in reasoning and logical fallacies and the doubtfulness of demonstration as about sensory illusion. Second, the formulation of the hyperbolical doubt that would have inspired the argument in the *Elements* if the *Discourse* influenced it is hyperbolical doubt in the form of the dream hypothesis. (The demon hypothesis is not mentioned in part 4 of the *Discourse.*) But Hobbes's argument about seemings and apparitions in the *Elements* is not connected to a dream hypothesis or even to the phenomenon of dreaming. Indeed, where dreams are mentioned in the *Elements,* in chapter 3, Hobbes takes the trouble to concede the skeptical point that there is no criterion for telling dreaming from waking, at least where the purported dream experience is in the past.[20] This concession would be out of place if a chapter earlier he had refuted a sweeping skeptical doubt geared to dream experience. Finally, it is extremely odd to base any account of the hyperbolical doubt on the highly compressed version of it given in the *Discourse.* Descartes himself treaded lightly in describing his metaphysical meditations there. As he said in his correspondence, he did not want to be taken for a skeptic, and he wished to avoid a metaphysical controversy that would delay further the already delayed appearance of his physics.[21] He organized the *Discourse and Essays* so that specimens of his scientific work would get the bulk of his readers' attention, and it was only when these came in for unexpectedly intense criticism, and when Jesuit teachers at La Flèche refused to endorse Descartes' science until they had seen more of its metaphysical basis, that he turned to a full-scale exposition of a first philosophy.

I shall now consider whether there is any evidence, apart from the passage quoted above from *The Elements of Law* or its counterpart in *Leviathan,* that Hobbes tried to answer, or even took seriously, the hyperbolical doubt. Perhaps there is, for when Hobbes himself states a first philosophy in *De corpore,* chapter 7, he makes use of a thought experiment that on the surface seems to resemble those conducted in the First Meditation, an experiment concerning an annihilated world. But this experiment not only differs from Descartes' in presupposing that the past is real: it is also based on the assumption that a physical world that used to exist caused

20. Hobbes 1969, 12 (pt. 1, chap. 3, sec. 10).
21. See AT 1:558.

the ideas from which first philosophy starts. In the *Meditations* Descartes does not assume from the outset that there is or was anything real beyond his own ideas: rather, the reality of everything is put into doubt in the First Meditation, and then the self, God, and mathematical natures are saved from the doubt, followed only finally by the physical world. The arguments that are supposed to show the skeptic that these things are real may not be very convincing, but that is beside the point, since what is at issue is how seriously Descartes takes the skeptic and whether Hobbes goes to comparable lengths not to beg the question against him. Interpreted as a passage in which Hobbes aspires to go as far as Descartes in the First Meditation to meet the skeptic on his own ground, the part of *De corpore* under discussion is a failure.

Perhaps, however, Hobbes never intended to go as far as Descartes. A number of passages suggest that Hobbes did not try to answer, and did not take seriously, the hyperbolical doubt. One of these comes from chapter 9 of *Leviathan*. Here Hobbes simply asserts that there is such a thing as knowledge of fact, this knowledge consisting of sense and memory, as if skepticism about perception and memory did not exist. There are remarks of a similar kind in part 1, chapter 6, article 1, of *The Elements of Law*. Finally, there are the Third Objections, which begin with comments on the First Meditation. Here Hobbes had the opportunity to write at some length about the hyperbolical doubt, and yet he produced only a couple of short paragraphs. His comments, brief and brisk, are those of a writer anxious to pass on to more interesting matters. It is clear from what he says that he saw nothing new in the dream and demon hypotheses—he dismisses the whole of the First Meditation as a stale discussion—and he seems to think that Descartes' difficulties in the First Meditation would have been solved by applying reasoning to the content of sense, as if the reliability of reason were not as much in doubt as the reliability of sense. The point is not only that Hobbes misses the message of the First Meditation; it is also that, contrary to what one would expect if Hobbes had engaged with skepticism, he does not seem to be interested in it.

It is true, of course, that close associates of Hobbes were concerned with skepticism and its refutation: Mersenne, for instance, and Gassendi. In *La vérité des sciences* Mersenne takes seriously skeptical arguments about physics and mathematics, and even accepts these arguments when they are directed against Aristotelian natural science; yet he suggests that these arguments are unsound when used against the new natural science. It is not clear, however, that Mersenne's antiskepticism rubbed off on Hobbes. Hobbes was undoubtedly a convert to modernity in science, but it was possible to be modern, for example, in the sense of believing that physical explanation is essentially geometrical, without having worked through, or responded seriously to, skepticism. Galileo is such a modern. And so, I would say, is Hobbes.

In the absence of firm evidence of Mersenne's having recruited Hobbes into an antiskeptical movement, another possibility is worth considering. It can be put by saying that Mersenne had a wide range of interests in the new science, that he encouraged work of all kinds in the new science for their own sake, and that the more rigorous pieces of this work, whether done with a view to refuting skepticism or not, were sometimes used by Mersenne himself as evidence of the weakness of skepticism about the mathematical sciences. According to this view, the purely scientific findings reported by members of Mersenne's circle would have been grist for Mersenne's antiskeptical mill. In other words, certain pieces of science—not only those produced by members of his circle but also the researches of such independent figures as Galileo—were open to appropriation by Mersenne as proof against the skeptics that real science is possible; but the pieces of science would not necessarily have been produced with anti-skeptical uses in mind. In particular, Hobbes may have developed scientific material on topics that were interesting to him in their own right, only to have them seized upon by Mersenne to further the arguments of *La vérité des sciences*. Thus, when Mersenne pressed Hobbes's *De cive* on Samuel Sorbiére to get him to reconsider his skepticism, he may have been appropriating work not written with antiskeptical intentions.[22]

Influences Closer to Home

To deny that Hobbes was spurred by Descartes' work or Mersenne's to devise an antiskeptical metaphysics is not to deny that both of them exercised some influence on Hobbes. It is clear that Hobbes engaged with the *Dioptrics,* and whatever the lost fifty-six-page letter said, we can assume that Descartes' work made an impression. It is hard, however, to go further. One commentator suggests that Hobbes may have been weaned off the Scholasticism of his own earliest work on optics, the *Short Tract,* by his exposure to the *Dioptrics,* but this is speculative in two ways.[23] Not only does no one appear to know whether any changes in Hobbes's optical theory are due to Descartes, but conjectures about his departures from

22. Popkin 1979, 147.
23. Sepper 1988, 530–31.

the *Short Tract* are further put into doubt by the possibility, first raised by Richard Tuck, that Hobbes did not write it.[24]

If it is difficult to show that Hobbes was influenced very much by the French intellectual milieu in his metaphysics and natural philosophy, it is perhaps even harder to show that his morals and politics were significantly affected. It is sometimes claimed that there is an antiskeptical motivation for at least Hobbes's moral philosophy and that this can be located, via a supposed Hobbesian reaction to Grotius, in a European skeptical crisis felt primarily in France. But as I have suggested elsewhere, the case for thinking of Hobbes either as a successor of Grotius or as one who was concerned to answer skepticism in his politics is pretty weak.[25]

The important influences on Hobbes's philosophical development and on his conception of his trilogy are probably not confined to French influences, still less to those that can be grouped together as aspects of a so-called skeptical crisis or a certain reaction to that crisis. Long before 1640, when Hobbes removed himself from England and began the most productive period of his intellectual life, there was a perhaps more important turning point. Sixteen twenty-eight marked the beginning of his transformation from a man of letters to a man of science. Temporarily out of the employment of the Cavendishes, Hobbes entered the service of Sir Gervase Clifton and was engaged as a companion to Clifton's son on his grand tour of the Continent. One of the best-known stories about Hobbes has its origins during this journey. Apparently in Geneva, in the library of some wealthy house, Hobbes looked for the first time at an open copy of Euclid's *Elements* and became absorbed in one of the proofs, which he followed step by step, with complete fascination. Aubrey writes that this episode made him fall in love with geometry. There is plenty of evidence in Hobbes's work that he regarded Euclid's book as one of the supreme models of scientific writing. Euclid is effusively praised in the preface of *De corpore,* and some of Euclid's results are openly incorporated into the mathematical sections of Hobbes's book. Perhaps also during the grand tour with Clifton, Hobbes was present at a discussion among some well-educated gentlemen about the nature of sense perception in which it emerged that none of the participants could say what sense perception is. This episode is significant because it probably put into Hobbes's mind the questions he would later consider in his correspondence with Descartes and in the second book of the trilogy.

After his return from the Continent in 1631, Hobbes went back into

24. Tuck 1988, 18.
25. Sorell 1993.

the service of the Cavendishes, as tutor to the young third earl of Devonshire. At about the same time he came into contact with the other Cavendishes, the earl of Newcastle and his brother Charles, who lived in Welbeck, close to the Devonshires' home at Hardwick Hall. It is possible that the gentlemen's discussion of the nature of sense involved them or took place at Welbeck rather than on the grand tour with Clifton. In any case, the Welbeck Cavendishes were interested in science, particularly its military applications. The earl of Newcastle sent Hobbes to London in the early 1630s to find a copy of a book of Galileo's. The earl's younger brother, Charles, had an even greater interest in science, and he acted as something of a patron and distributor of scientific writings, notably the writings of a scientist called Walter Warner. Robert Payne may have been asked to look at the writings that Charles Cavendish circulated, and Hobbes also commented on them at times. Cavendish had contacts with Mersenne and Mersenne's circle in Paris and can fairly be regarded as belonging to the scientific network that then existed in Europe. The smaller network that he ran closer to home, involving Warner, Payne, and Hobbes, may have as much of a claim to have prompted the ideas for natural science in the trilogy as Mersenne's circle in Paris. And other English aristocratic circles that involved Hobbes as a participant in political discussions have a much stronger claim than anything on the Continent to have inspired the political part of the trilogy. It is perfectly possible that in these English aristocratic circles, Hobbes not only discovered science but also acquired the habits of reading, writing, and talking about science that he drew upon as an objector to the *Meditations,* as a member of Mersenne's circle and the Parisian scientific community, and as the builder of the system that survives in his trilogy.

✤ 7 ✤

HOBBES VERSUS DESCARTES

I believe I am not the only reader of the *Objections and Replies* who has found the exchange between Hobbes and Descartes disappointing. When he writes against the *Meditations* in 1641, Hobbes is, in the eyes of the public, only the author of an English translation of Thucydides' *History of the Pelopponesian War* (1628).[1] That is a work of some interest, of course, though it is not a *Leviathan* (1651) or even a *De cive* (1642). Nevertheless, of all those who wrote objections to Descartes' work, it is Hobbes, certainly, who will become the philosopher whom history will regard as the most penetrating and original. From him we might have expected something illuminating, even brilliant, and we are disappointed. The author of the Third Objections is dogmatic; he insists on a materialist point of view without supplying any persuasive argument for it; and he seems not to understand the structure of the argument of the *Meditations,* imagining that the central argument Descartes gives for his dualism occurs in the Second Meditation. He seems not to notice the argument of the Sixth Meditation, and hence does not reply to it. In April 1641 Descartes writes to Mersenne: "I did not think myself obliged to deal more fully than I have in my responses to the Englishman, because his objections seemed to me so implausible that replying to them at greater length would have made them appear to be worth more than they are."[2] Can we, who know Hobbes better than Descartes knew him in 1641, reverse this rather severe judgment, which is also, up to a point, our own judgment?[3]

The original version of this essay was presented at the conference on the *Objections* and *Replies* held in Paris in October 1992 and appears in French in the proceedings of that conference. In translating the essay into English, I have made some modifications, mainly as the result of comments on the original version by my colleague Louis Loeb.

1. The subtitle of the French translation of the Third Objections will call Hobbes "a famous English philosopher," but this is a work of 1647, not of 1641.

2. AT 3:360.

3. At most Descartes knew Hobbes from their correspondence of 1641 on optics. Did he understand that the Englishman of this correspondence was the Englishman of the Third Objections? We may doubt it. The Englishman of the correspondence is clearly a materialist

Here we cannot abstain from judging whether Hobbes read Descartes well.

Let us look at the argument. In his second objection Hobbes cites the words of the Second Meditation, *sum res cogitans,* I am a thinking thing, and says, I agree. "From the fact that I think, or that I have a phantasm, whether waking or sleeping, it is inferred that I am thinking, for *I think* and *I am thinking* signify the same thing. From the fact that I am thinking, there follows *I am,* because what thinks is not nothing."[4] Up to this point Hobbes makes no objection. Perhaps here I am in disagreement with Monsieur Zarka, who seems to think that Hobbes admits the inference from *sum cogitans* to *ego sum* only "on the condition of noting that the verb *sum* in the proposition *ego sum* affirms existence in this instance." The implicit criticism would be that Descartes has confused "the logical function and the existential function of the verb *to be*" and that this confusion is essential to the Cartesian argument.[5] I do not find this confusion in Descartes, who does not employ the formula *sum cogitans, ergo sum* or even *cogito, ergo sum* at that point.[6] So I prefer to look for an interpretation of Hobbes according to which he does not accuse Descartes of having made that error.

Hobbes admits the inference from "I think" to "I exist," but he hesitates when Descartes adds, "that is to say, a mind *[mens],* a soul *[animus],* an intellect *[intellectus],* a reason *[ratio]*."[7] That seems to him a faulty inference, because it seems to identify the subject which thinks either with the act of the thing which thinks or with its very power of thinking. But all philosophers, Hobbes says, distinguish the subject both from its acts and from its powers. Descartes protests: yes, I have treated as identical the thing which thinks and the mind, or soul, but by these last words I did not understand a power of thinking or the act of thinking. I understood the thing which has this power or which performs this act. So I did not neglect an elementary distinction of the philosophers.

---

with respect to God (AT 3:287), whereas the Englishman of the Third Objections is very ambiguous on this issue, as we shall see. I do not find in the correspondence itself any certain indication that Descartes understood the identity of the author of the Third Objections. In 1643 he was able only to conjecture that the author of the Third Objections was also the author of *De cive* (AT 4:67). Armand Beaulieu (1990, 82) comments that Mersenne exercised a certain prudence in associating himself with Hobbes: "If it is only a question of scientific inquiries, everything is fine, as far as Mersenne is concerned. But if it is a matter of a philosophy which is too pagan, or some of whose reasonings are suspicious in the eyes of believers, it is better not to present too much those who inspire mistrust."

4. AT 7:192. See AT 7:27, l. 13.   6. AT 7:24, ll. 19–25, l. 13.
5. Zarka 1987, 144–45.   7. AT 7:27, ll. 13–14.

Should we find this response satisfactory? Certainly we must grant Descartes the right to explain himself, to say, I meant this, not that. But is there, perhaps, a reply which Hobbes could make to this reply? Why did Hobbes think Descartes meant something he did not mean? I conjecture that Hobbes had reasons to reject the hypothesis that Descartes understood by the term *mens* the subject which thinks. Hobbes could, it seems to me, say something like this: If by *mind* or *soul* you meant neither the act of thinking nor a power of thinking, but the very subject which thinks, either you conceive this subject as extended or you conceive it as nonextended. If you conceive it as extended, you conceive it as a body, and you grant everything I wanted. The thing which thinks is something corporeal. If you conceive it as nonextended, you conceive a thing which has a power (the power of thinking) but which has no other property more fundamental than this power, no property which could explain why the subject has this power. To conceive it in this way is to abandon the new philosophy, which requires that we explain every power of a subject by reference to some property which is more fundamental, but which is not merely a power. That would be to revert to the Scholastic way of speaking, as if we were to say: "The understanding understands; vision sees; the will wills; etc."[8] This is a mistake which Hobbes might well believe Descartes made, after having read the analysis of judgment in the Fourth Meditation, where Descartes seems to embrace the kind of faculty psychology Hobbes will criticize in *Leviathan*.[9] To avoid this error, it is necessary to conceive of the mind as a power of a subject which has some more fundamental property, a property which is not merely a power itself, and which can explain why the thing which thinks has the power to think. This property must be extension.

If we understand Hobbes' objection in this way, I believe we will see that Descartes' reply is not adequate. If I understand Descartes, when Hobbes accuses him of having begged the question in the Second Meditation, Descartes says that he did not claim, at that point of his argument, to prove that the mind was not something corporeal.[10] All he undertook to prove at that point was that it is conceivable that a mind should exist without a body, or rather that we can conceive clearly and distinctly a mind without a body. He proved the possibility of this clear and distinct conception when he found, after careful consideration, that he could not deny that he existed and that he thought, but that he could still deny that

8. AT 7:177.
9. AT 7:57. Hobbes' criticism is found in *Leviathan*, chap. 46, par. 28.
10. AT 7:175.

any body existed.[11] What he added, in the Sixth Meditation, was a proof that, if we understand something clearly and distinctly (as a complete thing), the existence of this thing is also really possible, as we conceive it, because there is a being who can do everything that we can conceive clearly and distinctly.[12] There is a God; this God is all powerful; an all-powerful being can do whatever we can conceive clearly and distinctly.[13] The most important property of God for the argument of the Sixth Meditation, which is where Descartes claims to establish the real distinction between mind and body, is his omnipotence, not his veracity. The Sixth Meditation adds real possibility to the purely conceptual possibility of the Second Meditation. And the real possibility that the mind exists without any body establishes that the mind is a being really distinct from any body, a conclusion which does not follow simply from the conceptual possibility without the proof of the existence of an all-powerful being.[14]

If I understand Hobbes, he understands that. His criticism is not that Descartes has not proven the real distinction in the Second Meditation but that in the Second Meditation Descartes has not proven even the conceptual possibility of a thing which thinks but is not extended. Hobbes' words, perhaps, are a bit ambiguous: "Therefore, it can be that the thinking thing is the subject of the mind, or reason, or intellect, and, therefore, is something corporeal. The contrary of this is assumed, not proven." [15] Which proposition is it whose contrary is assumed and not proven? Is it (a) The thing which thinks is the subject of the mind, and at the same time is something corporeal, or is it (b) If the thing which thinks is the subject of the mind, it follows that this thing is also something corporeal? The first proposition affirms a fact, and the contrary of a fact is another fact. The second affirms a necessity, of which the contrary is a possibility.

11. For Descartes, to have a clear and distinct conception of myself as an incorporeal thinking substance just is to find myself, after due reflection, compelled to affirm that I think and not compelled to affirm that I have a body (or indeed, that any body exists). I have discussed the Cartesian conception of clarity and distinctness in Curley 1986.

12. Cf. the First Replies, in AT 7:121.

13. AT 7:78, ll. 2–3.

14. This is an interpretation of Descartes that I have defended in Curley 1978, 193–99. To say that God can bring about whatever we can conceive clearly and distinctly as possible is not to say that he cannot bring about things that we cannot conceive clearly and distinctly as possible. Indeed, Descartes' doctrine of the creation of the eternal truths might suggest that God can bring about even things that we conceive clearly and distinctly to be impossible. I do not think it does imply that (cf. Curley 1984). But the important point is that the argument of the *Meditations* takes no position on this issue.

15. AT 7:173.

To prove a (conceptual) possibility, which is the project of the Second Meditation, is an easier task than proving a fact, which is the project of the Sixth.[16] Descartes thinks Hobbes is saying that he has not proven the contrary of (a), and he knows that he did not claim to prove the contrary of (a) until the Sixth Meditation. That is, he does not claim, in the Second Meditation, to have proven that the thing which thinks is really distinct from any extended thing. So he accuses Hobbes of not having understood the structure of the argument of the *Meditations*.

But I think Hobbes is saying that Descartes has not proven the contrary of (b), that is, that he has not proven that it does not follow from the fact that there is a thing which thinks (a subject of the act of thinking, or of the power of thought) that this thing is something corporeal. He has not proven that it is possible, conceptually, that the thing which thinks is something nonextended. If I am right, Hobbes has understood the structure of the argument very well. He does not grant that Descartes has proven even what he thinks he has proven in the Second Meditation, that is, the possibility of conceiving, clearly and distinctly, the existence of a thing which thinks without there being any extended thing.

I believe that the continuation of Hobbes' objection will confirm this interpretation:

> It is very certain that the knowledge of this proposition, *I exist*, depends on this, *I think*, as he has rightly taught us. But where do we get knowledge of this latter proposition, *I think*? Certainly not from anything but this: that we cannot conceive any act without its subject, such as dancing without one who dances, knowing without one who knows, thinking without one who thinks.
>
> And from this it seems to follow that the thing which thinks is something corporeal, for the subjects of all acts seem to be understood only as having a corporeal nature *[sub ratione corporea]* or as having the nature of matter *[sub ratione materiae]*, as he himself shows afterward in the example of the wax.[17]

Here we have something really surprising. Hobbes takes as an axiom, as it were, that we must conceive each subject of an act as corporeal or mate-

16. On my account of the matter, the distinction between the purely conceptual possibility established in the Second Meditation and the real possibility established in the Sixth is that real possibility requires that conditions sufficient for the realization of what is conceptually possible be actual. If God exists, then one element in a set of conditions sufficient for the existence of the mind without the body is real: there is a being with the power to bring about that existence. (If God chooses to exercise his power, then a set of sufficient conditions will be complete.)

17. AT 7:173.

rial (this is equivalent to our (b) above), and he claims that Descartes himself has shown the truth of this proposition by the example of the wax. Descartes is right, I think, to be astonished. Hobbes has advanced this position, Descartes says,

> without any reason, and contrary to all logic, and to every cus-tomary way of speaking . . . for the subjects of all acts are indeed understood as having the nature of substance [sub ratione sub-stantiae], or even, if you wish, as having the nature of matter [sub ratione materiae], viz., of metaphysical matter; but they are not on that account understood as having the nature of body.
>
> On the contrary, the logicians (and everyone else) are accus-tomed to say that some substances are spiritual and others corpo-real. Nor did I prove anything else by the example of the wax except that its color, hardness, and shape do not pertain to the formal nature of the wax itself [i.e., that one can conceive of ev-erything which necessarily is present in the wax without needing to think of those things]. Also, I did not speak at all there about the formal nature of the mind.[18]

Descartes is right, of course: he did not speak of the formal nature of the mind in his discussion of the wax; to say that we must conceive of each subject of an act as corporeal *is* to say something very unconventional; and if the only reason Hobbes offers for this proposition is that it follows from what Descartes has said about the wax, Hobbes owes us an explana-tion of this inference.

A man who tries to elevate his knowledge above what is common, Hobbes might say, paraphrasing Descartes himself, ought to be ashamed to rely on ordinary ways of speaking and thinking to provide arguments for his claims. That would be at least the beginning of a reply. But the question we must ask, I think, is this: Why did Hobbes think he could derive his contention that every subject is corporeal from the example of the wax? He understands the example of the wax to show that, although all the "acts" of the wax (its color, hardness, shape, etc.) were changed, we conceive it to be still the same thing because of what remains constant through this change. In the case of the wax, what remains constant is that it is an extended thing. And the position of Hobbes seems to be that, no matter what the subject is, what remains constant is the same thing: the subject is necessarily an extended thing.

But why this necessity? Descartes is right when he says that almost everyone denies that the subject is necessarily extended. He grants that

18. AT 7:175. The phrase in brackets is from the French version of the Third Replies.

we can speak of a matter which remains constant through the changes of a thing which thinks, but he insists that this matter is only a "metaphysical matter." The possession of this metaphysical matter does not imply that the thing is extended. Why say that it is (must be) extended?

Perhaps there are two reasons. In the first place, I believe that Hobbes would reject the notion of a metaphysical matter as a metaphor of the sort he thinks improper in philosophy.[19] The concept of matter, in Aristotelian philosophy, draws its meaning from the ordinary concept of matter as a stuff which can receive various forms. When we introduce the concept of a metaphysical matter, we say that it is like ordinary matter except that it is not extended. But when we do *that,* we eliminate everything which gave sense to the possibility of receiving various forms. So the concept of metaphysical matter is inadmissible in a rigorous philosophy.

But I think the situation is more complex than that. Hobbes seems to maintain that the alternative to his conception of the subject involves an infinite regress: "It is not by another thought that it is inferred that I think. For though someone can think that he has thought (a thought which is nothing but remembering), nevertheless, it is completely impossible to think that one is thinking (as it is to know that one knows). For this would lead to an infinite inquiry: how do you know that you know that you know that you know?"[20] We do not see easily why this objection is pertinent. Is it really not possible to think that one is thinking or to know that one knows? Why should someone always respond to the question "How do you know that you know?" Why does this endless inquiry arise?

Perhaps we will see some pertinence in Hobbes' objection if we imagine the following line of thought: The reason it is necessary to conceive a subject as corporeal is that no other property can serve as a principle of constancy through change. The Cartesian metaphysic supposes that thought in general can serve as a principle of constancy, as extension does; but there is a big difference between thought and extension. We can conceive of extension clearly without attributing to it any particular determination.[21] That is what we do in geometry, the only real science

19. Cf. *Leviathan,* chap. 5, par. 14. For *Leviathan* I recommend use of the edition I recently published with Hackett, which indicates important variations in the Latin *Leviathan* in the notes (Hobbes 1994).

20. AT 7:173.

21. In comments on my paper in Paris, Zarka denied that Hobbes would accept this, citing Hobbes's *Critique of Thomas White's "De mundo"* (chap. 27, par. 1) to show that for Hobbes there can be no clear conception of extension without a particular determination. It is not clear to me that the passage cited does show that, though perhaps other passages do.

we have at the moment. But we cannot conceive thought clearly without attributing to it some particular determination, some object of thought. This follows from the essential nature of thought, which, for Descartes as for Hobbes, is necessarily a representation of an object. Nevertheless, the thought which would serve as a principle of constancy through change cannot have any particular object. If it did, that object might change, and then it could no longer be a principle of constancy; but if it had no particular object, it would be incomplete, and so, again, it could not be a principle of constancy. If we imagine a sufficiently general object of this thought which is to be the principle of constancy (e.g., that I think), we can always ask, What is the object of this thought? So the infinite regress arises from the need to find a suitable object for the thought which is to serve as the principle of constancy. It must have some object, in order not to be incomplete; but if it is to be a genuine principle of constancy, it must have an object which is not too particular.

I admit that it is speculative to attribute such reasoning to Hobbes, but I believe that it can explain something which otherwise is very mysterious, namely, why Hobbes passes from his remarks about the wax to his accusation that Descartes cannot avoid an infinite regress in connection with thought. The weakness of my suggestion is that, if Hobbes did think something like this, he did not explain himself very well. But philosophy is a difficult subject, and if philosophers always explained themselves well, the history of philosophy would be a good deal easier than it is.

There is, of course, another argument in Hobbes for the inconceivability of a nonextended subject, which goes like this: to have an idea or conception of a thing is to have an image of it; but every image is necessarily the image of some extended thing; so, lacking an image of any nonextended thing, we cannot conceive it. I do not find this argument satisfactory. It is Hobbes at his most dogmatic. In the Sixth Meditation Descartes presents an argument for making a distinction between an idea and an image: we have an idea of a chiliagon which is completely distinct from the idea we have of any other polygon, such as a myriagon; but the image we have of a chiliagon is not distinct from the images we have of many other polygons; therefore, we must make a distinction between having an idea of something and having an image of it. It is one of the disappointing aspects of the exchange between Hobbes and Descartes that Hobbes does not reply to this argument. Perhaps Hobbes thinks he has replied to it in advance by what he has said about the two ideas of the sun in the Third Meditation (cf. his eighth objection). But he concedes there that in addition to the image of the sun derived from vision, we also have a different conception of the sun derived from reasoning. And I'm sure that Descartes would think that that gives him all he wants.

I pass now to the question of God. When Descartes writes that all logicians, and almost everyone else with them, are accustomed to say that among substances, some are spiritual and others corporeal, he is surely right. To say that no immaterial substance is even conceivable, as Hobbes has suggested in his second objection, is to say something very radical. It is also to say something dangerous, since this thesis seems to imply a rejection of the existence of that substance who is, in the tradition, the most important example of a spiritual substance. We do not know, perhaps, whether there is a God, or we do not know it before the arguments of the Third Meditation; but to say that the concept of an incorporeal substance implies a contradiction seems to lead to an a priori proof of the nonexistence of God.

Hobbes recognizes this danger in the Latin version of *Leviathan*. In the English version he had written that the complex expression *incorporeal substance* combines two names whose significations are contradictory.[22] In the third chapter of the appendix written for the Latin version of *Leviathan* (first published in 1668), he took the opportunity to reply to objections made to the English edition. These are *his* Objections and Replies. Writing in the form of a dialogue, he asks himself if the claim that *incorporeal substance* is contradictory implies that God does not exist or that God is a body.[23] His reply is that this is an affirmation that God is a body, a position he maintains is not heretical, since Tertullian maintained it, without being condemned by any general church council.

Nevertheless, Hobbes knows very well that affirming God's corporeality is, in the Anglican Church to which he claims loyalty, a heresy subject to excommunication. He has pointed this out himself in the first chapter of the Latin appendix.[24] And there are good reasons for this decree of the church, since it appears that if God were an extended substance, he would be divisible and hence could, in principle, suffer destruction. That is the argument Spinoza will seek to evade in Part 1 of his *Ethics*. There are also, I think, other theological problems in the affirmation that God is corporeal, which I have discussed elsewhere.[25] That, I sug-

22. *Leviathan*, chap. 4, par. 21.
23. Hobbes 1839–45, 3:561.
24. Ibid., 537.
25. See Curley 1992. A. P. Martinich has recently criticized this article in an appendix to *The Two Gods of "Leviathan"* (1992). Martinich's book is a sustained argument that Hobbes is not only a theist but an orthodox Christian (specifically, a Calvinist Anglican). I propose to reply to this in an article I call "Calvin and Hobbes."

gest, is why Hobbes is very circumspect about admitting this implication of his theory of substance when he writes something destined for publication. Before 1668 he had not written in any published work that God is corporeal.

Zarka has expressed a different opinion. He has written that it is necessary not to take literally this affirmation that God is corporeal.[26] To deduce from the fact that Hobbes said "God is corporeal" that he believed that God was corporeal would be "too hasty a conclusion, for it is also the constant doctrine of Hobbes that in his nature God is not knowable by reason. He affirms that God is corporeal only when his opponents press him to say something about God's nature . . . far from its being a rationally valid thesis, it is, instead, a blasphemy, the blasphemy of a reason which wants to pass beyond the limits of the knowable."[27] We are in agreement about one thing, at least: when Hobbes says something, we need not always conclude that he believes what he says. But we disagree about what it is that Hobbes said and didn't believe. Zarka takes as a strictly accurate reflection of his belief the words "God exists, but his nature is unknowable." I prefer a more radical interpretation: that Hobbes is an atheist, and that everything he says about God—that he exists, that he is unknowable, that he is a corporeal spirit—is a subterfuge, necessitated by the possibility of persecution.

How might one convince Zarka of this interpretation? Probably this would not be the work of a single afternoon. My suspicions come partly from a comparison of the critique of Scripture in *Leviathan* with that of Spinoza in the *Theological-Political Treatise,* and partly from the admiration Hobbes is reported to have expressed for the bolder work of Spinoza. If Aubrey's account is accurate, Hobbes, on reading the *Theological-Political Treatise,* said that he would not have dared to write so boldly, implying that he would have written something more along the lines of Spinoza had he dared. I take this to mean that had he dared, he would have openly rejected the whole theological tradition stemming from the Jewish and Christian Scriptures. To argue the case for this would require a very long discussion, which I have engaged in elsewhere.[28] Here let me limit myself to adding a few further observations.

26. Zarka 1987, 148.
27. Zarka 1984, 174.
28. See Curley 1992. I should perhaps add, since otherwise there is likely to be misunderstanding, that I do not think Spinoza was an atheist. The point of similarity with Hobbes is a rejection of the God of Abraham, Isaac, and Jacob. Spinoza does, I think, embrace quite sincerely the God of the philosophers, and since I see no reason why the theologians of the scriptural tradition should be granted a monopoly on the use of the term *God,* I would not

It is not strictly true that Hobbes affirms that God is corporeal only when his opponents press him to say something about God's nature. If we can believe the letter of Descartes of January 21, 1641, at any rate, Hobbes did affirm the corporeal nature of God in the letter to which Descartes was replying.[29] (But that is a letter, not a work destined for publication.) Also, it is not, strictly speaking, a matter of replying to a demand to say something about God's nature. To such a demand one could reply in the conventional way, by saying piously, We cannot grasp the nature of God. Rather, the problem is this: wishing to stamp out what he believes to be a dangerous superstition (the belief that there are immaterial beings), Hobbes has already said something which implies an answer to questions about God's nature; he has said that the term *immaterial substance* implies a contradiction, just as the term *round quadrilateral* does. He cannot avoid the implications of this thesis for the case of God. He must choose: if every subject is necessarily corporeal, and if incorporeal substances are a contradiction in terms, then either God does not exist or he is a corporeal substance. In what he writes, Hobbes chooses the latter alternative and resolutely disdains to discuss the theological difficulties such an affirmation involves. But one could hardly expect him to publicly embrace the former alternative, not in a period in which open atheism is a capital offense.

When he writes against the *Meditations,* Hobbes is more circumspect on the subject of God than he was in the correspondence on optics. Though what he writes concerning the Second Meditation might lead us to suspect that he will say that God is necessarily corporeal, what he in fact writes about angels permits us to think that he will allow the existence of some immaterial subjects:

> When someone thinks of an angel, sometimes the image of a flame presents itself to his mind, and sometimes that of a young child with wings; about this it seems to me that I am certain that it has no likeness to an angel, and therefore that it is not the idea of an angel. But *believing [credens]* that there are invisible and immaterial creatures, which are God's ministers, *we give [imponimus]* the thing believed in or supposed the name angel, though the idea under which I imagine an angel is composed of the ideas of visible things.[30]

class Spinoza as an atheist. But Hobbes is another matter. I see no sign in him of the rationalized religion that was so important to Spinoza.

29. AT 3:287.
30. AT 7:179–80.

There is a grammatical problem here, noted by Charles Adam and Paul Tannery and by William Molesworth. The participle *credens* is in the singular; the verb with which it should agree, *imponimus*, is in the plural. Faced with this problem, some translators translate in a way which produces an admission of the existence of immaterial creatures. Here is John Cottingham, for example: "I believe that there are invisible and immaterial creatures who serve God; and we give the name 'angel' to this thing which we believe in."[31] Other translators (e.g., E. S. Haldane and G. R. T. Ross) translate in a way that makes Hobbes merely give an account of a common opinion, without making it explicit that he shares that opinion. The grammatical problem permits a doubt about his intentions, which one can resolve in various ways.[32]

The case of angels is of great interest, since Hobbes says that the case of angels is like that of God. That is, we suppose that God, like the angels, is an immaterial being, and that is the reason why we cannot have an idea of God, since every idea or image is necessarily the idea of some corporeal thing. Does this "we" include Hobbes himself? We cannot know, but I think Hobbes' real position is quite different.

If God existed, Hobbes thinks, he would have to be a corporeal substance, since every subject of acts is necessarily corporeal. But if God were corporeal, he would not be God, for well-known reasons. Therefore, God does not exist. But it would not be prudent to say that, for reasons which are also well known. So it is necessary to find something else to say, and one says different things at differing times. Most of the time one says: God exists, but we cannot know his nature; so, we cannot know whether he is corporeal or not. This is normally the position, for example, of the English *Leviathan*.[33] But in proportion as there are reasons for thinking

---

31. CSM 2:127.

32. By the time of *Leviathan*, at least, Hobbes is unequivocal about rejecting the conception of angels as immaterial beings. In chapter 45, paragraph 8, he argues that there is no scriptural authority for that conception, it being a relic of paganism that has crept into Christianity.

33. I say "normally" because of an odd and interesting exception. In *Leviathan*, chapter 46, paragraph 12, as part of his argument that the schools of the Jews are unprofitable, Hobbes complains that they "turned the doctrine of their law into a phantastical kind of philosophy, concerning the incomprehensible nature of God and of spirits, which they compounded of the vain philosophy and theology of the Grecians, mingled with their own fancies, drawn from the obscurer places of the Scripture . . . and from the fabulous traditions of their ancestors." This anticipates Spinoza's criticism of Maimonides in the *Theological-Political Treatise* for reading Greek philosophy into Scripture; but it is surprising to see Hobbes criticizing Jewish philosophers for holding God's nature to be incomprehensible, when he had advocated the same position in *Leviathan* (chap. 31, par. 28).

that every subject is corporeal and that God, if he exists, is a subject, this is not really a satisfactory reply. Sometimes one says: Reason cannot instruct us on this subject, either one way or the other; therefore, it is necessary to recur to sacred Scripture, which tells us that God is immaterial. That is the position Hobbes takes in his *Critique of Thomas White's "De mundo."* But this is not a satisfactory position either. On the one hand, it seems that reason does lead to the conclusion that God is corporeal; on the other hand, it is difficult to find any texts of Scripture which employ the concept of an immaterial being. So sometimes, when one is old, and when one is publishing in a foreign language and in a foreign country, one says: God exists, and he is corporeal, and there is no theological problem with that position. This is the stance Hobbes takes in the Latin *Leviathan.* And that is as far as one can go in this period.

I conclude that Hobbes understood the structure of the argument of the *Meditations* better than we may have thought and that he makes an objection to the argument of the Second Meditation which attacks it where it is really vulnerable. When the question concerns God, though, he hesitates to draw all the logical conclusions of his argument, for reasons of prudence. His argument leads to an a priori proof of the nonexistence of God, and that is something one cannot say in that period, not if one wishes to die in bed.

VINCENT CARRAUD

$\mathcal{\check{y}}$ 8 $\mathcal{\mathclap{\check{c}}}$

ARNAULD: FROM OCKHAMISM
TO CARTESIANISM

Among the six sets of objections published in 1641 following the *Medita-tions*, those of Antoine Arnauld are notable, in several respects, so much so that they have acquired a unique status. First there is the judgment of Descartes himself, who writes of the Fourth Objections: "I consider them the best of all . . . because he has entered more than anyone else into the sense of what I wrote." [1] The Fourth Replies open with a rare compliment: "I could not wish a more perspicacious *[perspicacem]* or more obliging *[officiosum]* critic of my writings." Descartes praises not only Arnauld's kindness and civility but above all the care he has taken *(tam accurate,* "with such care"), so that "nothing has escaped him" *(nihil ejus aciem effugisse)*. [2] Moreover, the acuity of Arnauld's reading can be measured by another criterion, that of precision and of the technical difficulty of the questions raised (chiefly on the theory of ideas: completeness/adequation and material falsity), to the point that they require for Descartes himself a real effort of attention: "[I]t is extremely easy to be mistaken here—indeed, it could easily happen even to me: if I were looking at the senten-ces in isolation . . . , I would sometimes take one meaning for another." [3] In short, in contrast to some of the other objections, in particular those of Hobbes and Gassendi, Descartes is very well satisfied with the objec-tions of Arnauld.

That is why they are notable for another unique characteristic, which is the direct consequence of what I have already noted: Arnauld's re-marks are the only ones that gave rise to modifications and to what are in effect additions to the text itself of the *Meditations* (notably the *Sixth*) and of the First Replies. [4] Additions are especially conspicuous

---

1. To Mersenne, March 4, 1641, in AT 3:331, ll. 4–7.
2. AT 7:16–23.
3. AT 3:329, ll. 8–12. These lines doubtless apply to the Fourth Meditation in general.
4. One correction also bears on the Fourth Objections themselves, because of slightly faulty citation by Arnauld (omission of *quodammodo* in AT 7:108, ll. 22–23, then 109, l. 6, mentioned at 213, ll. 18–24). Descartes stresses the importance of this, however, since it

in the Synopsis.[5] Descartes himself wished the corrections and additions to appear as such, that is to say, between brackets.[6] Further, in this way Descartes makes of the special case of the Fourth Objections the model for all the objections he might wish for, the proof of his desire and his ability to correct himself in the search for truth. He also wrote to Mersenne on March 18, 1641: "I ask you to change the following things in my *Metaphysics,* in order in this way to let it be known that I have deferred to his judgment, and so that others, seeing how ready I am to take advice, will tell me more frankly what grounds they have against me, if any, and be less obstinate in contradicting me without grounds."[7] The corrections due to the Fourth Objections thus carry out the resolve repeated several times by Descartes: "I requested expressly in the *Discourse on Method* of all those who would find errors in my writings, to deign to let me know of them; and I announced loudly that I was entirely willing to correct them."[8] Thus the objections of Arnauld alone fulfilled the dialogic ideal that Descartes had held to since the *Discourse,* in providing him with the opportunity of proving that pertinent remarks or objections had a real and measurable effect on his thought and thus on the very text of the *Meditations.*

Finally, the objections of Arnauld are plainly notable in that they are not set a priori against the Cartesian project, insofar as they do not oppose to it the global coherence of a philosophy already constituted, such as the Thomism of the First Objections, or in the course of being constituted, such as that of Hobbes or of Gassendi. Doubtless one could say of the latter objections that, because they have their own coherence, which it is essential to relate to the philosophy of their author, they sound so much the more like a dialogue of the deaf, and give no evidence of a reading of the Cartesian arguments themselves that is "attentive and free of all prejudgments." From this point of view, Arnauld's youth at the time of his reading of the *Meditations* was doubtless a nonnegligible quality (Baillet emphasizes it), in that it perhaps permitted a kind of philosophical

allows him to attenuate the incongruity inherent in the expression *causa sui.* See the letter to Mersenne of March 18, 1641, in AT 3:337, l. 4–338, l. 5.

5. See AT 7:15, ll. 6–12, on the distinction between the evil and the false in the Fourth Meditation. To this addition we owe an extremely interesting *hapax legomenon* in the *Meditations,* namely *speculativus.*

6. AT 3:335, ll. 5–6.

7. AT 3:334, ll. 3–9.

8. To Mersenne, Aug. 30, 1640, in AT 3:169. See the *Discourse,* in AT 6:75, l. 21–76, l. 5.

innocence that made him able to read Descartes without prejudice—or at least with favorable prejudices.[9] Arnauld writes to Mersenne in sending him his objections: "You have known for a long time in what esteem I hold his person, and how highly I rate his intelligence and learning."[10] Arnauld was doubtless already favorably inclined toward Cartesianism at the time he received the *Meditations,* if only through his reading of the *Discourse* and the *Essays.*

As Henri Gouhier insists, the objective rapprochements between Descartes and St. Augustine contributed not a little to the seduction of Arnauld; they have an essential share in explaining "the attitude taken spontaneously [I insist on this 'spontaneously'] by the young doctor in the face of the *Cartesian phenomenon.*"[11] None of the four references to St. Augustine furnished by Arnauld goes to support an objection. On the contrary, they were to assist Descartes when he was attacked by theologians. As all the commentators repeat, Arnauld's Augustinianism made him very sensible to the similarities between Descartes and the bishop of Hippo. Thus it is important not to underestimate this initial reason for his adherence to Cartesianism. On the other hand, it is very possible that the philosophical relevance of Cartesianism furnished him with a new occasion to read St. Augustine, in whose work several passages now shone in a new light, in particular some beyond the texts on grace that were customarily cited. This is the case precisely for the works to which Arnauld refers in the Fourth Objections and which, it seems, are not those Saint-Cyran had keenly recommended that he read: *De libero arbitrio, De quantitate animae, Soliloquia,* and *De utilitate credendi.*[12] Thus Cartesianism may have reactivated an interest in St. Augustine the philosopher and not merely the doctor of grace. In particular, one can measure the displacements in the Augustinian corpus itself effected by the successive appearance of new quotations and new rapprochements. I concur here with the thesis proposed by Geneviève Rodis-Lewis: "Descartes, far from being in a position to receive the echo of an ambient philosophical Augustinianism, would have allowed his contemporaries to rediscover the

9. Baillet insists on the youth of Arnauld, not less remarkable than that of Pascal, which had astonished Descartes a little earlier (1691, 2:125).

10. AT 11-A:153. I consider this "for a long time" below; see n. 41.

11. Gouhier 1978, 123. On Descartes and Arnauld, see esp. chap. 2, secs. 1 and 3, and chap. 5, secs. 1 and 2.

12. See, in Arnauld 1775–83, *La vie de Messire Antoine Arnauld* by Noël de Larrière and the correspondence with Saint-Cyran, in vol. 1 and the preface to vol. 10, esp. p. v. Saint-Cyran even gave Arnauld a list of texts of St. Augustine on grace (10:ii, with a reference in Quesnel in *L'histoire abrégée de la vie de M. Arnauld*).

metaphysical originality of the thinker of Hippo, in his affinities with the Cartesian program." [13] This would no doubt be true for Arnauld himself, before it came to hold par excellence for Malebranche.

However that may be, Sainte-Beuve summarizes in a formula the very nature of Arnauld's objections: "Arnauld, in fact, understands Descartes more than he combats him." He continues: "[A]dmirable logical mind, he will not be a discoverer in philosophy, and if only his theology be satisfied, he will willingly cleave to the new master." [14] Francisque Bouillier agrees: "It is a happy accident for Descartes to encounter such an adversary, let us rather say, such a disciple." [15] It is true that Arnauld appears in the *Objections,* at least methodologically, as Cartesian as Descartes himself: on several occasions, it is precisely in the name of Cartesian principles that he opposes Descartes, bringing to light what seem to him to constitute internal contradictions. For example, when Descartes maintains at the same time that falsity resides only in judgments and that there are ideas that are materially false, Arnauld concludes: "quod ab *eius* principiis dissonum mihi videtur," "which seems to me inconsistent with his own principles"; or again, the notion that the *esse objectivum positivum,* the positive objective being, of an idea could come from nothing "praecipua *Clarissimi Viri* fundamenta convellit," "overthrows the distinguished gentleman's chief foundations." [16] In short, in the Fourth Objections, Arnauld is no less accomplice than objector.

From then on, Arnauld will take a firm stand on Cartesian positions, over against Descartes himself, in what the editors of his *Oeuvres* have called his New Objections (the two letters of June and July 1648), against Father Etienne Le Moine in his "Ecrit contre l'anti-cartésien," and above all against Malebranche and Leibniz. [17] During the half century that opens

13. Rodis-Lewis 1990, 101 (p. 104: "L'augustinisme métaphysique leur était ainsi révélé"). See also the analysis of the successive references to rapprochements, in particular book 10 of the *De Trinitate* mentioned by Arnauld only in 1648: "L'apparition successive de ces rapprochements montre déjà combien la familiarité des contemporains de Descartes avec les écrits du saint était limitée" (102f.). See also Alquié 1974, 187.

14. *Port-Royal,* bk. 6, chap. 5; Sainte-Beuve 1930, 5:351. For the manner in which Descartes treats Arnauld, Sainte-Beuve relies on a remark of Leibniz on reading *La vie de Monsieur Des-Cartes* by Baillet. Sainte-Beuve's pages on Arnauld's Cartesianism conclude as follows: "Well then! one can make of Arnauld a great logician, one can make of him a Cartesian disciple, and the first among the disciples; one can never make him a philosopher" (449).

15. Bouillier 1868, 1:223.

16. AT 7:206, l. 10, and 207, ll. 23–24.

17. For the New Objections, see Arnauld 1775–83, vol. 38, following the Fourth Objections. For the "Ecrit contre l'anti-cartésien," see *Examen d'un écrit qui a pour titre: Traité de l'essence des corps, et de l'union de l'âme avec le corps, contre la philosophie de M.*

up before him, he who is perhaps the first Cartesian will also be the most exact Cartesian, the most rigorous, and the most loyal—but I offer this point for discussion to Steven Nadler. What is it that leads the theologian Arnauld on two occasions to return to the practice of philosophy? When he could attack directly the theology of Malebranche's *Traité de la nature et de la grâce,* it is, says Henri Gouhier, "the desire to reestablish the genuine Cartesian theory of ideas" that leads him to make the "philosophical detour." Then again, in the discussion of 1686–87 with Leibniz, "through his pen, it is Descartes who is resisting the author of the *Discourse on Metaphysics,* posing as evident the definition of the essence of matter by extension and not even conceiving of the possibility of *modernizing* substantial forms." [18] Thus on these two well-known occasions, it is the defense of orthodox Cartesianism that required Arnauld to do philosophy.

Arnauld's union of Cartesianism and Augustinianism has been amply commented on (and in particular his theory of ideas, which constituted "his" philosophy—as it still does in the recent book by Nadler), and all the historians of philosophy have rightly considered the Fourth Replies as an important, indeed inaugural moment for the Cartesianism of Arnauld himself—no doubt because of certain acute questions of Cartesian thought, and the new and decisive formulations about ideas, the concept of *causa sui* or the eucharistic physics Arnauld compels Descartes to state. Through this, finally, the Fourth Replies are important for the history of rapprochements between Descartes and St. Augustine. [19] But a decisive, even an inaugural, moment does not necessarily constitute an absolute point of departure, even for Arnauld himself. To my knowledge, none of the commentators on Arnauld mention what might have been prior to the Fourth Objections, except for that theological "Augustinianism," philosophically indeterminate but always presupposed, and often recalled by Arnauld himself, who places it back before 1635 (the date of his *Tentative*). They write in this way, presumably, in order to insist on the idea according to which Arnauld's authentic Augustinianism owes nothing to the reading of Jansenius, which underlines a unique domain, theological and ecclesiological (and not philosophical), since it bears in a common-

---

Descartes, in Arnauld 1775–83, 38:89f. The New Objections are presented thematically, like the first, with the addition of questions *De re quanta a locali extensione non distincta* and *De vacuo.* See also Baillet 1691, 2:347–48, where the identity of the objector is not mentioned.

18. Gouhier 1976, 126–27. See also Rodis-Lewis 1973, 389.

19. See chiefly Verga 1972; Senofonte 1989; Nadler 1989; and Ndiaye 1991.

place manner on questions about original sin, about grace, and about predestination.[20]

During the winter of 1640–41 (and more generally during these two years), however, Arnauld was not content with just writing his objections. On the contrary, in the letter to Mersenne that introduces them, he mentions many other activities that keep him much occupied *(molestis occupationibus)* and the difficulty of finding "the tranquility and the leisure" required for "a very deep meditation and a great withdrawal of the mind into itself."[21] Hence in the project of attempting to understand the coherence of the Fourth Objections and to grasp how Arnauld was able to become so rapidly and so exactly Cartesian, we may perhaps find it not wholly unprofitable to consider what occupied him at the time, especially his course in philosophy.[22]

Arnauld in 1641–42

Arnauld had on hand a copy of the *Meditations* and of the First Objections and Replies by December 1640 at the latest. Descartes received his objections in February 1641.[23] Descartes apologized on March 4 for not yet having sent his reply to Arnauld, and again on March 18 and 31 (the delay was without doubt due to the work imposed on Descartes by the

20. See, for example, letter 415, to Du Vaucel: "It is no great marvel to have found in St. Augustine, after forty years of study, the same principles that Jansenius expounded, since I in fact found them in 1635, when I sustained my tentative, four or five years before the publication of the book of M. d'Ypres" (Arnauld 1775–83, 2:243). See also Besoigne 1752, vol. 5, bk. 11, p. 341: "He sustained it six years before the book of Jansenius appeared, before he ever knew there was any man on earth called Jansenius"; and Quesnel 1697, 24. The *Augustinus* was published in 1640.

21. AT 9-A:153; AT 7:197, ll. 9–13.

22. Henri Gouhier is the only person to mention Arnauld's instruction, but he merely rests content with a conjecture: "No more than [St. Augustine] did Arnauld consider philosophy useless. He taught it at the college of Mans in 1638–1640 and 1640–1641; he could appreciate the insufficiency of the physics of the Schools. We can appreciate the pleasure he takes in reading a copy of the *Meditations,* coming shortly after this experience" (1976, 126).

23. The letter to Mersenne of January 28, 1641, mentions the last objections received, those of Hobbes (AT 3:293, ll. 16–19). As for Arnauld, he may have read the *Meditations* before December 1641, if one considers with Bernard Rochot that "the importance of Arnauld's contribution alone prevented its being joined to the Second Objections, which remained anonymous, and which we find today of a different spirit" (Mersenne 1932–88, 10:357).

question of the Eucharist).[24] Some matters of chronology are necessary to situate that decisive winter of 1640–41 in the life of Arnauld, since the winter of the Fourth Objections and Replies falls within the two years of the course in philosophy that Arnauld was giving at the Collège du Mans, a branch of the Sorbonne, and they took place while he was engaged in his own degree curriculum. Thus I shall be concentrating on some months that were absolutely decisive.

1. On November 14, 1635, Arnauld defended his *Tentative,* through which he earned his bachelor's degree. This "tentative," dedicated to the General Assembly of the Clergy, offered Arnauld the occasion to develop the doctrine of St. Augustine on predestination and grace, in particular by asserting that the omnipotent will of God is the cause of the predestination of the saints. This being so, he placed himself publicly (after having done so in private) in opposition to the professor whose lectures he had been following since 1633, Father Lescot, an influential man, confessor of Cardinal Richelieu, and of whom it is "claimed that the notes that he dictated were only an abridgement of the theology of Vasquez": this point is essential.[25] Jansenist historiographers interpret this conflict as that between Scholasticism and patristics, or between speculative theology and positive theology, and see in it the effect on the young Arnauld of Saint-Cyran, who had advised him "to add to the notebooks of his Professor the works of Saint Augustine."[26] The initial friendship of Lescot, to whose lectures Arnauld had taken a dislike, changed to hatred, and it is probable that this situation had a bearing on the refusal of Richelieu to see Arnauld enter into the Society of the Sorbonne.[27]

2. The degree curriculum required the defense of four theses, which took place on the following dates: November 12, 1638, *Theses pro Sorbonica, de Christo, et de antiqua et nova lege;* 2) November 21, 1639, *Theses pro minore ordinaria, de ecclesia;* 3) January 13, 1640, *Theses pro majore ordinaria, de sacramentis;* and finally, 4) on December 18, 1641, *Theses theologicae pro Actu Vesperiarum* ("Quid finis legis?" "What Is the End of the Law?" on Romans 10). The doctor's bonnet was

24. To Mersenne, March 4, 1641: "I am not yet sending you my reply to M. Arnauld, partly because I have been otherwise occupied, and partly because I do not want to hurry; but I believe, all the same, that I can send them to you in a week's time" (AT 3:328, ll. 24–28). See also the letters of March 18, in AT 3:334f., and of March 31, in AT 3:349f.

25. See Arnauld 1775–83, preface to vol. 10, esp. pp. ii–iii.

26. Ibid. Arnauld received permission to correspond with Saint-Cyran at the end of 1638, and then later to go to visit him at the chateau of Vincennes. See Arnauld 1775–83, 10:v.

27. On Lescot's influence on Cardinal Richelieu and the consequences that resulted for Arnauld, including his exclusion from the House and Faculty of the Sorbonne because of the censure of 1656, see Quesnel 1697, 24–26.

assumed the day after the last defense, which for Arnauld would have been December 19, 1641.[28] Moreover, I emphasize this point: the Fourth Objections were written while Arnauld was preparing for his degree. He was not yet doctor of theology of the Sorbonne.

3. Nor was he yet a priest, but in the course of preparation for the priesthood (he was to be ordained during the Ember Days of September 1641). However, in 1639–41 he wrote *On Frequent Communion*.[29] The work was completed before he was priest and doctor.[30] It would not be published until April 1643. It is certainly not unprofitable to take note that Arnauld's scruples about the compatibility of the Cartesian theory of sensible qualities and extended substance with the church's teaching on the Eucharist are exactly contemporaneous with the composition of his great thesis on the Eucharist, although the question here is one exclusively of positive theology and never of eucharistic physics.

4. However, the treatise *On Frequent Communion* is not the only book that Arnauld wrote during these two extremely productive years. In addition to polemical works and editions,[31] Arnauld wrote *The Necessity of Faith in Jesus Christ,* in which he restated, developed, and amplified his first theses *pro Sorbonica* and proved especially that he was not ignorant of the philosophy of his time, since he attacked La Mothe le Vayer and contemporary Neostoicism on the question of the salvation of pagan philosophers.[32]

5. Last but not least, these two years were those during which Arnauld fulfilled the obligation of every future degree-holder in theology to teach philosophy. Arnauld's intellectual qualities, manifest from the time of his *Tentative,* did not gain him exemption from the course in philosophy, but it did gain him the waiver of "the formality that prescribes that the course

28. For the text of these theses, see Arnauld 1775–83, 10:9–32; see also 1:12f.; and Quesnel 1697, 37f. For Descartes' opinion of Arnauld, with or without the doctor's bonnet, see the letter to Gibieuf of January 19, 1642: "[A]lthough it is not long that M. Arnauld has had the doctorate, I still esteem his judgment more than that of half the ancients" (AT 3:473, ll. 10–12).

29. "He was already working on the book *On Frequent Communion,* and M. de Saint-Cyran, who considered his work an excellent preparation for the priesthood, urged him to continue this work" (Arnauld 1775–83, 1:14).

30. Ibid., 16: "The book *On Frequent Communion* was finished before he had received the Doctorate."

31. In 1639 he published the *Apologie pour M. l'Abbé de Saint-Cyran,* and in 1641, the *Extraits de quelques erreurs d'Antoine Sirmond, jésuite* and the *Dissertation sur le Commandement d'aimer Dieu, contre Antoine Sirmond, jésuite* (Arnauld 1775–83, 29: nos. i and ii). Also in 1641 he edited the *Peregrinus Hiericontinus* of Florent Conrius (10:lxxxvii).

32. *The Necessity of Faith* was published in 1701; see Arnauld 1775–83, 10:iii. See also Gouhier 1987a, 126–28. *De la vertu des païens* was published at the beginning of 1641.

must be completed before entering on the degree course." Thus he began two years of teaching philosophy in October 1639.[33]

Hence the winter of 1640–41, that of the Fourth Objections, found Arnauld occupied in writing two books and in pursuing the preparation of his licenciate at the same time that he was teaching a course in philosophy. All this, at the very least, suffices to justify the occupations mentioned at the beginning of his letter to Mersenne. I will therefore examine the course in philosophy that he taught and see whether there is a relation between that course and the objections, which are contemporary with it.

The course in philosophy was traditionally divided into four parts—logic, ethics, physics, and metaphysics; mathematics normally belonged to another branch of instruction.[34] What we know about Arnauld's course concerns theses from these five subjects. However, I shall leave aside what concerns mathematics. Arnauld's opposition to the Scholastic philosophy—if the singular has any sense—in favor of a positive theology established by the Fathers clearly did not free him from the duty of devoting himself to exercises that were by definition Scholastic. We know Arnauld's course through the conclusions that were drawn from it and that constituted the material of the defense of the *Tentative* of Charles Wallon de Beaupuis. Wallon de Beaupuis was one of Arnauld's students, and it is he whom Arnauld was to send to visit Descartes at the time of his stay in Paris in July 1644, when Arnauld could not go himself.[35] Now we know that it is Arnauld himself who edited these *Philosophical Conclusions,* at whose defense he presided on July 25, 1641, at the Sorbonne—that is to say, exactly one week before the

33. Arnauld 1775–83, 1:12–13. In fact, he had to teach a course in philosophy in order to belong to the Society of the Sorbonne. Arnauld applied for this for the first time on the eve of Pentecost in 1641. On the other hand, he had been admitted as early as Easter of 1638 (the beginning of his course for the degree) to the *hospitality* of the House of the Sorbonne. See Quesnel 1697, 20; and Besoigne 1752, vol. 5, bk. 11, p. 350. All his biographers insist on the difficulty of combining the "exercises for the degree" and the teaching of philosophy; see, e.g., Quesnel 1697, 38.

34. On the organization of the *rationes studiorum,* especially in the Jesuit colleges, see Gilson 1976, 117–19; and Brockliss 1987, esp. 185f. and 337f.

35. See Baillet 1691, 2:129: "Knowing that M. Descartes was in Paris during the summer of the year 1644, he could not resist sending one of his friends, a young ecclesiastic, to visit him [marginal note: M. Wallon de Beaupuis], and having him offer his services." Sainte-Beuve reports that "Arnauld and Descartes never saw one another except through him" (*Port-Royal,* bk. 4, chap. 4, in Sainte-Beuve 1930, 3:567–68). We find "Walon" as well as "Wallon" de Beaupuis.

presentation of the *Meditations* to the Faculty of Theology of the Sorbonne on the first of August.[36] Thus these theses hold very great interest for the historian of philosophy.

P. Quesnel, the biographer of Arnauld, insists on the originality of Arnauld's course, in which he was not content to repeat without discrimination various theses of the School: "[D]uring his licenciate, he had to compose a course of philosophy, and to teach it publicly. And this work, which is enough to occupy other men entirely, was so much the greater for M. Arnauld, since he was not a man to copy someone else's writings, nor to embrace opinions that he had not meditated on and examined with care."[37] Thus Arnauld's course appears original and well considered. Another piece of evidence is even more interesting. In a letter on December 13, 1642, Mersenne assured Voetius, already for some time in search of criticisms of every kind addressable to Cartesianism, that Arnauld had found nothing that did not satisfy him in Descartes' explanations.[38] This letter was translated by Claude Clerselier in the second volume of Descartes' letters, published in 1659; Arnauld would have had many occasions to deny this if he had thought it inexact. And Baillet quotes the same passage:

> I recently asked [the word *recently* is an addition to the original Latin—in any case this situates the letter about a year before the end of the course] the Author of the Fourth Objections, who is considered one of the subtlest philosophers, and one of the greatest theologians of this Faculty, if he had no rejoinder to the replies that had been made to him by M. Descartes. He answered that he did not have any, and that he considered himself fully

36. See Arnauld 1775–83, 38:i: "M. Arnauld composed the theses in philosophy, which we here give to the public, before he had finished his degree; and at the end of the course of philosophy at the College of le Mans, which he was obliged to teach in order to be received by the House and Society of the Sorbonne, having been admitted up to then only to what is called simple *hospitality*." These *Conclusions* are reproduced in volume 38 of the *Oeuvres* of Arnauld (pp. 1–6). See also *La vie* (1:13–18). The notice of the *Conclusions* ends with this indication: "De his, Deo duce, et praeside Antonio Arnauld, in sacra Facultate Licenciato Sorbonico, et Philosophiae Professore respondebit Carolus Walon Bellovacus. Die 25 Julii, quae sacra est S. Jacobo, an. Dom. 1641, à I. ad vesperam. In Aula Coenomanensi."

I owe the confirmation of the August 1 date to Jean-Robert Armogathe, who was able to find the record in the register of the Sorbonne preserved in the National Archives in Paris, and thus to identify the person who reported on the *Metaphysics* of Descartes, Jérôme Bachelier.

37. Quesnel 1697, 38.

38. On the whole dossier (the five letters of Voetius to Mersenne), and on Voetius' misgivings, see Mersenne 1932–88, vol. 11. See also Verbeek 1988.

satisfied. He even added that *he had taught and publicly defended the same philosophy [ipseque docuisset et publice sustinuisset eandem philosophiam]*; that it had been vigorously attacked in full assembly by several learned men, but that it had not been overthrown or even shaken.[39]

At the close of his biography, Baillet wants to show that "Descartes does not plagiarize from anyone." To do this, he passes in review a certain number of proper names and comes to Arnauld, "who is the only one living today, of all those who could boast of having anticipated M. Descartes in anything. This celebrated doctor has always appeared far from believing that our philosopher had ever been in the situation of borrowing anything from him, although he had publicly taught in the University of Paris the same philosophy as that of M. Descartes, before the latter had published the first Essays of his philosophy." [40] As we see it, the paraphrase of the same letter is as clear as possible: Arnauld taught the same philosophy as Descartes, before the *Meditations*. It can only be a question of the course of 1639–41, of which we possess the conclusions. Do these theses actually teach the same philosophy as Descartes? Is it a question of what we understand by Cartesianism? Arnauld had then read only the *Discourse* and the *Essays* (as the Fourth Objections reveal), and probably some letters that Mersenne had had circulated, and he was engaged in reading the *Meditations* and proposing objections to them.[41] Could he give a Cartesian course? In what sense can one say of the conclusions of Arnauld's course that they express the same philosophy as that of Descartes?

Thus the text of the *Conclusions* is not only important; it is essential for understanding the philosophy of the young Arnauld and, beyond the special case of Arnauld, for grasping exactly in what the novelty of Cartesianism consisted and to what extent it could be found *admissible* in the Sorbonne around 1640.

39. Baillet 1691, 2:128; emphasis added. The Latin text is given in Mersenne 1932–88, 11:375, ll. 44–52; and AT 3:603. For the French translation, see Descartes 1936–63, 5:239–40n.

40. Baillet 1691, 2:544. Baillet gives a marginal note to the same letter of Mersenne, but paraphrases it this time and adds the last very explicit proposition. Remember that *La vie* was published in 1691 and that Arnauld died in 1694. Does this not increase further the value we can attribute to this passage and to Mersenne's letter itself?

41. The Fourth Objections refer to the *Discourse on Method* on several occasions (AT 7:199, l. 1; 207, l. 1; 215, l. 2). See the *jam pridem* of the opening (197, l. 6). See also Baillet 1691, 2:25: Arnauld, "having formerly read with pleasure the *Essays* of the *Method* of M. Descartes, . . . acquiesced in the wish of Father Mersenne with the hope of finding the same pleasure in the reading of the *Meditations*."

Arnauld and Ockhamism

For present-day readers of Descartes, the *Philosophical Conclusions* could not pass straight off for a Cartesian text, even for a text of Cartesian Scholasticism, such as existed in the Netherlands in the 1640s.[42] The whole of the part on "conclusions on ethics" rests on an initial Augustinian problematic, in the opposition of supreme good to beatitude. Thus *beatitude* is defined as "the transformation of man into God, that is, the perfect and sempiternal union of the whole man with the highest good."[43] We find here adherence to the Augustinian supreme good.[44] Nothing in Arnauld's course rests on the Thomist definition of *blessedness* (which rejects the concept of transformation in favor of participation) or even on the problematic of the opposition of a natural end and a supernatural beatitude. That the conclusions on ethics are of Augustinian origin is not surprising, but this point will have to be modified as a result of our reading of other theses, particularly those concerning freedom.[45] However—this is the chief point—the conclusions on logic, on physics, and on metaphysics turn out to be Ockhamist.

1. The overall Ockhamism of the theses in logic is indubitable. I shall

42. The moral theses invoke Epicurean pleasure, Stoic uprightness, and Peripatetic virtue, before opposing, in classic fashion, the Platonists to the philosophical sects: "Saniores Platonici, qui nonnisi divini luminis participatione, et in altera demum vita beari posse hominem agnoverunt. Sed ut viderint illi quo eundem esset, non viderunt qua, et per multos errores a veritatis tramite recesserunt."

43. All quotations of the *Philosophical Conclusions* are from Arnauld 1775–83, *38*, 1–6.

44. See, for example, the *De Trinitate* 8.3.5, or the *Enarrationes in Psalmos*, in particular pp. 49 and 66, or again the *Sermones* 47 and 166. For a global presentation of the Augustinian doctrine of beatitude, see the *Dictionnaire de théologie catholique*, cols. 504–8, and, still more generally, the *Dictionnaire de spiritualité*, s.v. "Divinisation."

45. The theses in fact offer a definition of freedom of which certain elements, admittedly not very original, are very close to an Ockhamist formulation. Point 6 begins in classic fashion by distinguishing freedom with respect to means and freedom with respect to the end: "Hominem ex homine tollit, qui liberum negat. Humanae vero libertatis naturam in eo recte statuas, quod homo praecedente cognitione in finem ut finis est, fertur, et media quae accommodata fini videntur, eligit. Hinc duplex libertatis actio, circam finem, et circa media; sed potissima quae circa finem." "He who denies freedom removes man from man. Indeed, it is established that the nature of human freedom consists in the fact that man acts toward the end insofar as it is the end, and chooses the means that are appropriate to the end. Hence the double action of freedom, toward the end, toward the means; but the most important is that toward the end." But then Arnauld links freedom and indifference in almost precisely Ockhamist terms: "Libertatem indifferentia plerumque comitatur, nequaquam tamen ad illam plane necessaria." "Indifference usually accompanies freedom, but nevertheless without being absolutely necessary to it." Next he repeats an essential Ockhamist affirmation: "Qui peccare non potest, sine dubitatione liberior quam qui potest." "The

present several examples, which are clearly Ockhamist in their content as well as in their formulation and which bear on theses essential to Ockham's thought.[46] First, the nature of the singularity or the univocal function of the sign: "There is no true community of things according to nature. Whatever is, is singular and something through itself [quidem per seipsum]. What therefore is universal? Either the sign, which, without varying, personally and immediately represents many singulars; or many singular natures represented by a common notion and a univocal name." Still more distinctly Ockhamist, if that is possible, is: "What is more particular than particularity itself?"[47] Second, in the same unit on logic, the definition of relation is equally Ockhamist: "Relation is nothing beyond its basis and its term: nothing not reciprocal; it is properly terminated neither to the absolute nor to the relative." ("Relatio nihil est praeter fundamentum et terminum: nulla non mutua; nec ad absolutum, nec ad relativum proprie terminatur.") Last but not least, it is again in logic that we find the famous pronouncement of the univocity of being with respect to God and the creature, of Scotist origin—in logic rather than in metaphysics, hence according to the Ockhamist reinterpretation of Scotus: "Being belongs synonymously to God and the creature, to substance and accident."[48]

Now a most remarkable event is associated with the defense of this thesis, which Quesnel reports in his *Justification* of Arnauld:

> M. de la Barde, a very learned man and subtle theologian . . . attacked this proposition and pushed the respondent [Wallon de Beaupuis] vigorously. The President [M. Arnauld], seeing him embarrassed in the thick of the difficulty, came to his rescue; but he himself found himself so hard pressed, and so entirely con-

---

one who cannot sin is without doubt freer than the one who can." Ockham says, for example, "Posse peccare nec est libertas nec pars libertatis, sed magis minuit libertatem" (*Questiones . . . in IV Sententiarum Libros* 1, dist. 10, q. 2 K; 4, q. 14; see Baudry 1958, 137, and 136, pt. 4, on freedom and indifference). We can already see very well how these two last formulations will be susceptible of a Cartesian sense, after the reading of the Fourth Meditation (AT 7:58).

46. For all the comparisons that follow, I have used Baudry 1958; the texts cited in Vignaux 1934; William of Ockham 1988; Baudry 1950; and Alféry 1989. I thank Olivier Boulnois for confirming these first comparisons.

47. And what follows: "Facile tamen potest rationis opera fieri universalis, in Dialecticae leges peccant qui vel individuo vago, vel cuivis termino quamtumvis communi, dum in propositione singulari praedicatur, universalitatem tribunt."

48. The assertion of the univocity of being is itself found initially in John Duns Scotus, *Ordinatio* 1, dist. 3, par. 25f. See Scotus 1988, 94f. This thesis continues: "Unica proinde categoria sufficit, quae omnia complectatur, etiam illum qui super omnia. Multae ex Aristotelicis Categoriis sola ratione distinguuntur."

vinced by the arguments of the disputant, that he believed he must pay homage to the truth. He preferred to admit that he had no further reply to make, rather than to search for subterfuges and pretexts such as professors never lack on such an occasion, and that he could have found better than many others. I believe, Monsieur, that you are right, he said to M. de la Barde; and I promise you that from now on I abandon my opinion to follow yours.[49]

Jansenist historiographers did not fail to report this event, extraordinary for a defense: the abandonment of a thesis for its contrary, not only by the candidate, but by the director himself, in the face of the rigor and tenacity of objections, and to see in it a preeminent virtue of Arnauld, testifying at the same time to his humility and to his love of truth. But there remains a basic question. How could someone like Arnauld change his mind in the midst of a defense on a thesis so massive, so determining, and so historically decisive? I am tempted (this is only a conjecture) to see in it a supplementary indication of his Ockhamism. For if the Scotists hold to this thesis with as much conviction and defend it with as much relentlessness as the Thomists exhibit in holding to the equivocity of the concept of being, only an Ockhamist can know its formulation to be so indeterminate that he can abandon it, since the concept thus attributed is in fact nothing *in re*, nothing "from the side of the thing."[50] For the rest, in conformity to Arnauld's promises, it is indeed the equivocity of being that was to be defended by Wallon de Beaupuis six years later, on the occasion of his own defense in theology, where the theses were again directed by Arnauld: "Nothing belongs univocally to God and to creatures."[51]

49. Quesnel, *Discours historique et apologétique,* at the head of the *Justification* of Arnauld, p. 24. See Arnauld 1775–83, 38:i–ii. That the adversary who succeeded in making Arnauld change his mind was the Oratorian Léonor de la Barde is not perhaps a matter of indifference, since he was considered by Descartes to be one of his most zealous Parisian defenders, along with Gibieuf, also of the Oratory. See the letter to Gibieuf of Jan. 19, 1642, in AT 3:472, l. 2–473, l. 7. La Barde is also one of the objectors of the Sixth Objections (see the letter to Mersenne of June 23, 1641, in AT 3:385, ll. 15–24, and to the abbé de Launay, possibly on July 22, 1641, in AT 3:420f). The three-sided relations of Arnauld (Wallon de Beaupuis), La Barde, and Descartes would merit investigation.

50. See Ockham, *Opera theologica,* 1 *sent.,* d. 2, q. 9, in particular, and the commentary on it by Boulnois: "Ockham completes the univocist movement begun by Duns Scotus, and separates in a decisive manner the domain of the concept from that of the thing: the fact that a concept is univocally predicable of a thing does not imply any formal identity or real composition in that thing" (in Scotus 1988, 35–36).

51. Arnauld 1775–83, 10:33. Notice, however, that the relation of substance to accident disappears with the announcement of the thesis, as well as what concerns the predica-

2. The theses on physics are no less Ockhamist than those in logic. Moreover, Ockham is cited by name with respect to the lack of distinction between extension and the extended thing: "Ockham was of the opinion that extension is probably not really [re] distinguished from extended things." Other definitions are Ockhamist, for example those of action— "Action is not an entity superadded to the term"—of the continuum divisible to infinity, or of place and of surface.[52]

3. Let us go on to the theses of metaphysics, which are much briefer. The same statement clearly obtains. To begin with, the definition of metaphysics is Ockhamist-Scotist: "Metaphysics is the science of being." So is the distinction between essence and existence by the mind alone: "Only the mind distinguishes essence from existence." The two propositions that follow seem again to confirm this: "The source of the possibility of things must not be sought anywhere but in the immense power of God. All the positive properties of being, which are not distinct from the real thing, are nonetheless objects of examination."

This inquiry thus permits an irrefutable conclusion: the *Philosophical Conclusions* defended by Wallon de Beaupuis and directed by Arnauld are Ockhamist. Thus the course in philosophy given by Arnauld in 1639–41 was also Ockhamist. But this conclusion requires the formulation of an accompanying hypothesis. The hypothesis I should like to submit to your judgment is that the propositions of Arnauld exhibit a kind of pre-Cartesianism. In fact, a certain number of Ockhamist formulations are susceptible of receiving a Cartesian sense. Everything happens as if the Ockhamist pronouncements of Arnauld were in some way themselves

---

ments. But above all it appears henceforth in theology, and thus acquires a metaphysical, and no longer a logical, status. May one not suppose further that the reading of the *Principia* and the assertion of the equivocity of substance (finite/infinite) was not without effect on the confirmation of the initial abandonment of univocity in 1641?

52. "IV. Locus est *superficies, etc.* Operam ludunt, qui corporum locum internum, vel ab inanitate, vel ab immensitate divina repetere satagunt. Locus externa res est, et si quis internus statuendus esset, non alius sane quam propria corporum extensio. Vacuum naturae odio est; imaginaria spatia, rationi: nec in illis Deus, nec universum in loco. Ut possit esse corpus in vacuo tamquam in loco communi, non tamen ut in proprio. Vulgaris Philosophorum *ubi* nulla necessitas."

"IV. Locus est *superficies* etc. [Aristotle, *Physics* 4.4]. They are wasting their time, those who are contented with repeating that the internal place of bodies comes either from the void, or from divine immensity. Place is an external thing, and if one could suppose one that was internal, it would be nothing other than the extension itself of the body in question. Nature abhors a vacuum; reason, imaginary spaces: God is no more in them than the universe is in a place. If it were possible that body should be in the void, it would be as if in a common place, and not in its own place. The *ubi* of vulgar philosophers has no necessity."

Cartesianizable. We have seen that Arnauld had read the *Discourse* and the *Essays* well before writing his objections. Now a number of Ockhamist pronouncements in physics are—and were to be—capable of acquiring a Cartesian meaning, to begin with, as far as concerns the definition of extension already cited ("Extension is not to be really distinguished from the extended thing").[53] In the same way it is plain to see how the Cartesian doctrine, as early as the close of the first discourse of the *Meteors,* will be able to confer a renewed sense and coherence on a proposition such as "The continuum divisible to infinity is not made of atoms, nor does it even have any parts distinct from the divided part in the thing itself." The same holds, again, for what follows: "There is and can be no actual infinite except God."[54]

As for the theses of metaphysics, they can be read as making a clearly anti-Suárezian assertion, insofar as it is opposed to the theory of the independence of the eternal truths: "[To imagine] real essence outside God for all eternity is to dream while awake." Further, we find there, along with the thesis already cited, and more Ockhamist than Ockham-like, a proposition that leaves open the possibility of the doctrine of the creation of the eternal truths: "The possibility of things is to be sought nowhere except in the immense power of God." Should one then not speak of a pre-Cartesian or Cartesianizable Ockhamism?

A last point is highly significant for the theoretical ambivalence of the *Conclusions.* It belongs to the second part of the theses in metaphysics.[55] The following proposition occurs there (again anti-Thomist, since it af-

53. See the most developed formulations in the *Principles* 2, arts. 11 and 12, in AT 8:46, l. 1–47, l. 5, after the *Rules,* Rule 14, in AT 10:441, l. 4–443, l. 29. See also the letter to Mersenne of June 23, 1641, which Arnauld could have known, in AT 3:387, l. 15–388, l. 2. In the *Examen* Arnauld will support the Cartesian definition of extension with that of the Fathers, in particular St. Augustine, by citing *De Genesi ad literam* 7.21 ("Corpus dicimus naturam quamlibet longitudine, latitudine et profunditate spatium loci occupantem") and Question 5 of the *83 Questions:* when the Fathers "spoke as philosophers, that is to say, when they considered bodies according to the natural notions that we have of them, that the essence of the nature of bodies was to be extended, and that they could not be without extension" (Arnauld 1775–83, 38:105; also 113).

54. See the *Meteors,* in AT 6:238, l. 28–239, l. 12. If it is not forbidden to align the definition of place in Rule 14 with *Principles* 2, arts. 10 and 12, it is perhaps not illegitimate to ask what it owes to the second discourse of the *Dioptrics,* in AT 6:102, l. 12. It is the same for rarefaction. The proposition maintained there ("Non fit rarefactio per novam corporum extensionem") perhaps owes something to the close of the first discourse of the *Meteors* (AT 6:238); in any case it seems to anticipate the *Principles* 2, arts. 7–8, in AT 8-A:43, l. 27–45, l. 7.

55. In fact it is essentially composed of a discussion of the metaphysical determination of the hypostasis, where Arnauld takes his place in a sharp debate that opposes him to Cajetan,

firms the self-evidence of the existence of God): "To one who is scrupu-
lously attentive, and who is free from the prejudices of opinions, it is no
less evident that God exists, than it is that two is an even number." The
expression *diligenter attendenti,* "to one who attends diligently," occurs
everywhere in Descartes, just as the requisite of "being free from the prejudices of opinions" is in fact very frequent there.[56] But above all, we recognize the exact textual repetition of the Second Replies (5th question):
"eritque ipsis *non minus per se notum, quam numerum binarium esse
parem.*"[57] It is true that the analogy between the self-evidence of the
existence of God and that of two as an even number is not rare.
Nevertheless, the textual exactness of the rapprochement is striking:
the theses of Ockhamist origin in Arnauld's course purely and simply
assimilate a postulate of the Second Replies—arranged in geometrical
fashion.

Conclusion

It seems to me indisputable that the first philosophy of Arnauld is Ockhamist, at least that this is true of his course of 1639–41, which displays
an academic, or Scholastic, philosophical position (in effect an anti-
Thomist position) contemporary with the *Meditations.* Thus we can see
very well how it was to be possible, in the 1640s, for Ockhamist propositions (perhaps already in conjunction with certain Cartesian theses of the
Essays accompanying the *Discourse on Method* or of the letters to Mersenne) to be read or rethought as Cartesian, and thus how they could

Capreolus, or Suárez: "II. Modo aliquo a se distincto sibique superaddito substantiam non
indigere, quo fiat Hypostasis, regia via est, quam Patres nostri calcaverunt, ideoque tutis-
sima et tenenda. Per modalem illam subsistentiam divinae Incarnationis mysterium obscura-
tur potius, quam illustratur." "II. That substance has no need of a certain mode different
from itself, added to itself, by which it becomes hypostasis, is the royal road that our Fathers
trod, and for that it is the surest and the one to be held to. The mystery of the divine
Incarnation is rather obscured than illuminated by this modal subsistence."

56. See in particular the Second Replies: "ejusque conclusio per se nota esse potest iis qui
a praejudiciis sunt liberi, ut dictum est postulato quinto" (AT 7:167, ll. 5–7).

57. AT 7:163, l. 28–164, l. 1; emphasis added. The same proposition is repeated once
more in 1647 in the *Quaestio theologica* defended by Wallon de Beaupuis, immediately
after the affirmation according to which nothing belongs univocally to God and to the creature: "Porro ex eo solo quod utcumque Deum concipiamus, seu quod aliqua in nobis entis
perfecti idea sit, clare demonstrari potest Deum existere. *Imo attente res expendenti, et
a sensuum praejudiciis libero, non minus per se notum est Deum esse, quam numerum
binarium esse parem*" (Arnauld 1775–83, 10:33; emphasis added).

easily be made Cartesian. In this sense, we can understand that Arnauld's Ockhamism can be considered as pre-Cartesian. On my view, only this pre-Cartesian Ockhamism could justify Arnauld's having declared in 1642 that he had given a course of the same philosophy as Descartes: that is to say, a course including academic statements, perhaps influenced by the *Essays* of 1637, or even *in extremis* (the theses in metaphysics) by the *Meditations* and the First and Second Replies themselves, sufficiently flexible, in the confused Scholastic field of the years 1630–40, to be able to acquire a Cartesian meaning.

Thus in a certain way what happens in the case of Ockham is analogous to what happens in the case of St. Augustine. Arnauld was Augustinian in theology, that is to say, on matters of grace, of predestination, and in particular of justification, but also quite ordinarily.[58] Descartes permitted him the reading of another St. Augustine and conferred on this underestimated Augustinian corpus its metaphysical standing.[59] Thus the citations of Augustine that Arnauld proposes to Descartes in the Fourth Objections accredit Cartesian thought as much as they acquire from that use a new dignity and a new coherence: a properly epistemological coherence, like the tripartite *to understand / to believe / to opine* of the *De utiliate credendi,* for example, or the distinction of imagination and thought or intelligence according to the *De quantitate animae* and the *Soliloquia;* and metaphysical coherence, like the new status of the Augustinian *cogito* according to the *De libero arbitrio.*[60] In short, if St. Augustine lends credit to the *new* Cartesian philosophy, it is indeed that new philosophy that makes it possible to reread him and to rediscover what is specifically epistemological or metaphysical in him. Descartes makes it possible to re-

58. Predestination and grace are the subjects of the *Tentative* of November 14, 1635.

59. See Rodis-Lewis 1990, esp. 115: "If it is impossible not to appreciate the vigor of theological Augustinianism in the time of Descartes, the essential object of the Saint's works could mask, for those who knew them best, the originality of his philosophical argumentation. It would be improbable that Arnauld would not have read the *De Trinitate* before 1648, but at the moment when he discovers the *Meditations* the rapprochement has not yet imposed itself on his mind."

60. See AT 7:216, l. 11–217, l. 5; AT 7:205, l. 13–206, l. 28; and AT 7:197, l. 24–198, l. 8. The reference to the *De utilitate credendi,* chapter 11, complicates the Cartesian problematic, since it makes it pass from a bipartition to a tripartition. Doing this, it restores (perhaps without knowing it) Rule 12 of the *Regulae* and the three types of *impulsus* that were defined there *(phantasia, libertas propria,* and *potentia superior),* thus furnishing a possible source for this last text (AT 10:424, ll. 1–18). Be that as it may, this objective encounter of Descartes and St. Augustine is a perfect example of the double movement described here.

assimilate Augustinian passages that had been philosophically neglected or underestimated. In this double sense, St. Augustine prepared the way for Descartes.[61]

I believe that what happens with Arnauld's Ockhamism is analogous. Ockhamism was without doubt the academic expression of Arnauld's philosophy, in part through anti-Thomism, in part also because it was easier to reconcile with his initial theological Augustinianism. The ambivalence of the Ockhamist propositions that we have discovered also permits Cartesianism to reassimilate them by giving them a new and decisive sense and coherence. Common, and for that very reason ambiguous, statements thus find themselves philosophically reinvested, and this reinvestiture is called Descartes. But as distinct from his Augustinianism—which has become a Cartesianized Augustinianism[62]—Arnauld's Ockhamism, precisely because it is academic and philosophical, had no reason to survive his Cartesianism, or to survive *in* his Cartesianism, which was acquired thanks to the Fourth Objections and was thenceforth to be definitive.[63]

61. Consider the celebrated formula of Clauberg in the index of the *De cognitione Dei et nostri,* referring back to the *exercitationes* 4, par. 23, and 66, par. 7 (Clauberg [1691] 1968, 2:601, 704): "Augustinus cartesianae Metaphysicae favet." "Augustine is well disposed to Cartesian metaphysics."

62. See Gouhier 1977. Gouhier discusses the categories of Cartesianized Augustinianism and Augustinized Cartesianism. Arnauld (like Fénelon) is perhaps not a philosopher, but if he has a philosophy, it is that of Descartes.

63. Witnesses to this are the polemic with Father Le Moine and then all the double debate with Malebranche and Leibniz, where Arnauld, veritable Descartes *redivivus* in the face of post-Cartesian theses, will once more give evidence of an exemplary acuity and exactitude.

STEVEN NADLER

∂ 9 ∾

OCCASIONALISM AND THE QUESTION
OF ARNAULD'S CARTESIANISM

The fourth set of objections to Descartes' *Meditations*, composed by Antoine Arnauld (1612–94), has the distinction of being the only set submitted by someone who would himself go on to become a committed
Cartesian. But just what kind of Cartesian did the author of the Fourth
Objections become? Arnauld is usually regarded as an uncritical
Cartesian, as one of Descartes' more rigorous but unimaginative disciples
in the seventeenth century. His contribution to the history of Cartesianism tends to be seen more in the application of his keen analytic skills to
an orthodox defense of that philosophy than in any innovations that
would lead to its further development. This general impression is certainly
not without foundation. Arnauld is unequivocal in his admiration for
Descartes' philosophy, and unlike some other, more deviant Cartesians
(e.g., Nicolas Malebranche), he rarely criticizes Descartes explicitly.[1] Even
his objections to the *Meditations* stand in contrast to the other sets of
objections by their generally constructive spirit. Arnauld is not attacking
Descartes in the objections; rather, he is playing within a system to which
he is strongly attracted. This is evidenced by the fact that almost all of
Arnauld's criticisms are directed at Descartes' argumentation and not at
his theses. Finally, Arnauld's loyal and vociferous public defense of Descartes later in the century against the attacks of the church, the state,
and the universities served only to undermine his own already precarious
standing with those authorities (and, much to their chagrin, that of his
Jansenist associates at Port-Royal as well).

It is, however, a mistake to think that Arnauld is wholly uncritical in
his attachment to Cartesian metaphysics. In fact, if Malebranche and

---

1. On the rare occasion when he does criticize Descartes, it is usually on theological, not
philosophical, grounds. For example, Arnauld notes of Descartes that "his letters are full
of Pelagianism and, besides these points of which he was convinced by his philosophy—
such as the existence of God and the immortality of the soul—all that can be said of him
to his greatest advantage is that he always seemed to submit to the church" (letter 243
to ?, Oct. 18, 1669, in Arnauld 1775–83, 1:671).

Géraud de Cordemoy are to be regarded as "unorthodox" Cartesians for the extensive occasionalist (and, in Cordemoy's case, atomist) elements they introduce into their mentor's system, then (although for slightly different, less extreme reasons) so must Arnauld.[2] Arnauld's originality lies in the fact that he quite early perceives a problem peculiar to Descartes' dualism—what we now call a mind-body problem—according to which real interaction between two such radically different substances is inconceivable. Moreover, contrary to the usual textbook mythology on occasionalism, Arnauld is the only Cartesian both to recognize such a problem and to use occasionalism to resolve it. Acknowledging this complicating feature of Arnauld's relationship to Descartes' philosophy is important not just for revising our view of Arnauld as a philosopher but also for understanding the development of occasionalism and Cartesianism in the seventeenth century.

In the first section of this essay, I will briefly examine Arnauld's general attitude toward Descartes' mind-body dualism. In the second section, I will consider various aspects of Arnauld's limited occasionalism and the problems he sees in dualism that, in part, motivate his use of that causal doctrine. In the third section, I will relate Arnauld's occasionalism to that of other prominent Cartesians and discuss the significance of Arnauld in the development of Cartesian metaphysics.

Arnauld and Descartes' Dualism

In Arnauld's eyes, the most advantageous and important element of Descartes' system—that which, I believe, constitutes its greatest attraction for him—is its dualism of mind and matter (or body). In the Cartesian ontology, these two distinct substances have absolutely nothing in common besides their substantiality—neither in essential nature, as the kinds of substances they are, nor in the properties or modifications of which they are capable. Mind is unextended thinking and is modified by particular thoughts; body is thoughtless extension and is modified by shape, size, divisibility, and mobility.[3] For Arnauld, this radical separation of mind and body is of crucial value on both philosophical and theological grounds, above all in that it represents the only means of truly securing

2. There are already occasionalist elements in Descartes' system regarding the motion of inanimate bodies; see *Principles of Philosophy* 2.36–42. For a discussion of this, see Garber 1993. But these occasionalist elements by no means add up to the full-blown systematic occasionalism explicitly argued for by Malebranche, Cordemoy, and others. See also Prost 1907.

3. See Descartes, Second and Sixth Meditations; and *Principles of Philosophy* 1.53, 2.4.

the independence of the soul from the perishable body and, hence, of proving its immortality. In a long work dedicated to defending Descartes' views on the separation of mind and body, Arnauld insists that "M. Descartes [has] established the distinction of the soul from the body extremely well, which is the only solid foundation of its immortality. . . . If there is anything for which M. Descartes should be commended, it is for having so well separated our soul from our body, and having so well established that these are two totally distinct substances, of which only one is material."[4] While Arnauld remained unwaveringly committed to this dualism, he was not unaware of the potential problems for such an ontology.[5] One problem, which Arnauld acknowledges, concerns interaction. Another potential problem concerns what might appear to be a reduction of the body to a mere instrument or vehicle of the soul. Arnauld himself raises this question in his objections to the *Meditations:* "It seems . . . that the argument [of the Second Meditation] proves too much, and takes us back to the Platonic view (which M. Descartes nonetheless rejects) that nothing corporeal belongs to our essence, so that man is merely a rational soul and the body merely a vehicle for the soul—a view which gives rise to the definition of man as 'a soul which makes use of a body'."[6]

Some forty years later, Arnauld—by then a devout Cartesian—defends Descartes against just this charge, using exactly the same analogy that Descartes employs in the Sixth Meditation:

> By drawing our attention to the mutual and natural correspondence of the thoughts of the soul with traces in the brain, and emotions in the soul with movements of the spirits, the philosophy of M. Descartes suffices to convince us that our soul is not to our body as a pilot is to his ship; but that these two parts are united together in a union so great and so intimate, that they together make but a single whole, which is all that reason and Christian doctrine requires us to believe about the union of soul and body.[7]

Arnauld must have been satisfied by Descartes' initial response on this matter in the Fourth Replies, since he does not raise it again in the second

---

4. *Examen d'un écrit qui a pour titre: Traité de l'essence du corps et de l'union de l'âme avec le corps, contre la philosophie de M. Descartes* (hereinafter cited as *Examen*), in Arnauld 1775–83, 38:138.

5. Dualism informs all of Arnauld's writings. See, for example, *Des vraies et des fausses idées,* chap. 5, and his letter to Leibniz of September 28, 1686.

6. AT 7:203; CSM 2:143.

7. *Examen,* in Arnauld 1775–83, 38:141. See Descartes, Sixth Meditation, in AT 7:81; CSM 2:56.

set of objections he sends to Descartes in a letter in 1648. In fact, he begins that letter by noting that "what you have taught about the distinction of the mind from the body seems to me quite clear, perspicuous, and divine."[8]

Arnauld and Occasionalism

Occasionalism, in its strong version, is the doctrine that finite created beings—minds and bodies—have no causal efficacy whatsoever and that God alone is a true causal agent. Malebranche, Cordemoy, and Arnold Geulincx are strong, thoroughgoing occasionalists. They employ God's unique causal activity in all three contexts in which questions about natural causation arise: body-body, mind-body, and mind alone. Arnauld's occasionalism, on the other hand, is more limited, both in its scope and in its claims about causal powers. He uses the doctrine to explain only mind-body relations. Moreover, as we shall see, he is still willing to grant to finite substances real causal efficacy. Let us consider the matter on a case-by-case basis.

In the case of body-body relations, there are no reasons for believing that Arnauld is an occasionalist and good reasons for thinking that he is not. In none of the contexts in which Arnauld takes an occasionalist stand does he mention body-body relations, and in none of the contexts in which he discusses body-body relations does he ever mention God as a universal or primary or efficient cause of the motion of bodies (although God is naturally referred to as the "first" cause of motion). In fact, Arnauld positively claims that one body in motion is the real cause of the motion of another body with which it comes into contact. While demonstrating that God is the cause of sensible perceptions in the soul on the occasion of bodily motions in the brain, Arnauld argues against attributing to bodies any power to bring about mental effects: "For the motion of a body [cannot] have any real effect other than moving another body." He adds that "one ought not to have recourse to the First Cause unnecessarily," and he apparently does not see in the body-body case, unlike the mind-body case, any necessity for such a divine recourse.[9]

This is an odd position for a Cartesian as sophisticated and as familiar with the literature of his Cartesian contemporaries as Arnauld.[10] Arnauld

---

8. Arnauld to Descartes, June 1648, in Arnauld 1775–83, 38:68.

9. *Examen*, in Arnauld 1775–83, 38:146–47.

10. For example, Arnauld knows Malebranche's arguments against granting motive force to bodies; see *Réflexions philosophiques et théologiques sur le nouveau système de la nature et de la grâce* (hereinafter cited as *Réflexions*), in Arnauld 1775–83, 39:232.

accepts the Cartesian picture of matter as identical with pure extension. He explicitly rejects attributing to inanimate bodies substantial forms, occult virtues, or any other active or spiritual properties. For example, in the *Examen* he insists that all material events and effects can be explained "mechanistically, that is, by the disposition, configuration, and movement of their parts, or of the fluids that run through an infinity of small perceptible and imperceptible channels . . . without recourse to powers, faculties, and virtues, which are only indeterminate words by which . . . philosophers disguise their ignorance."[11] (A Cartesian body, then, should be entirely passive and inert, devoid of force and capable only of receiving motion, not of causing or sustaining it. And surely Arnauld recognizes this, as is suggested by his objection to Leibniz' doctrine of spontaneity: Arnauld writes that "what seems to me very hard to understand [is] that a body that has no motion can give itself motion."[12] Arnauld also accepts the doctrine of continuous creation or divine sustenance that grounds both the body-body occasionalisms of Malebranche and Louis de La Forge[13]—whereby the motion of a body just is its being re-created in different successive relative places from one moment to the next—and the occasionalist account of the motion of inanimate bodies in Descartes.[14] Where, then, is there room for bodies to be real causes of motion?

This is not the place to go into Arnauld's views on physics. But briefly, what I suspect he has in mind is a world in which bodies qua parcels of extension, while not endowed with active motor force, nonetheless (and

11. *Examen,* in Arnauld 1775–83, 38:151–53. See also his letters to Leibniz of September 28, 1686, March 4, 1687, and August 28, 1687, in which he argues energetically against Leibniz' views on substance, which to Arnauld appear to resuscitate the substantial forms of the Scholastics.

12. Arnauld to Leibniz, Aug. 28, 1687, in Leibniz 1965, 2:106. See also *Des vraies et des fausses idées,* in Arnauld 1775–83, 38:212: "Matter is not moved of itself, or through itself, because it does not give itself motion."

13. See Malebranche, *Entretiens sur la métaphysique,* Dialogue 7; and La Forge 1974, chap. 16.

14. See Descartes, *Principles of Philosophy* 2.36–42. On Arnauld, see *Des vraies et des fausses idées,* in Arnauld 1775–83, 38:284: "Creatures [are] powerless to subsist for a single moment unless they are sustained by a kind of continual creation by the same hand that drew them from nothing in order to give them being." And yet Arnauld explicitly divorces the divine sustenance or continuous creation doctrine from its apparent occasionalist ramifications. He insists that to give a substance a new modification—which is all moving a body involves—is not the same as creating or conserving the substance with this modification. These are two distinct activities. See *Dissertation sur la manière dont Dieu a fait les fréquents miracles de l'Ancienne Loi par le ministre des Anges* (hereinafter cited as *Dissertation*), in Arnauld 1775–83, 38:689–90. Garber (1987) discusses such a distinction in Descartes as a difference between being a modal cause and a sustaining cause.

perhaps with divine concurrence) collide and transfer among themselves
the motion that God as the first cause of motion originally implanted in
the world.[15] He still lacks an account of force (which occasionalist
Cartesians identify with the will of God), and he would still need to explain how the motion of one body, being a modification of that body's
substance, could be transferred to another body, since a mode cannot be
communicated from the substance it modifies to another. But these are
matters to which Arnauld did not give a lot of attention. Arnauld would
probably consider such an account simply a faithful rendering of Descartes' views on the matter, as he might have interpreted them from the
Principles of Philosophy.[16]

In the case of body-mind relations, particularly in the explanation of
how sensory effects result in the mind from motions communicated to
the brain via the nerves from an external object, the answer is also clear:
Arnauld favors an occasionalist account. The real efficient cause of the
sensations is not the bodily motions that ordinarily precede them, nor
is it the mind itself. Rather, God directly and immediately causes those
sensations in the mind on the occasion of the appropriate motions in the
brain, as determined by the laws of mind-body union.

The corporeal motions in the sense organs and the brain do not cause
sensations, Arnauld argues, because the motion of a body can have no
real effect other than moving another body. The motion of a material
substance certainly cannot cause any effect in a spiritual substance, since
a mind is by its nature incapable of being moved or pushed.[17] Arnauld is
refusing to introduce here the distinction (conveniently employed by La
Forge and others to account for body-mind interaction) between univocal
and equivocal causation. A univocal cause brings about in another object
some effect similar to what exists in the cause (for example, one body in
motion causes motion in another body), while an equivocal cause brings
about in another thing an effect unlike anything in the cause (as motions
in the body cause thoughts in the mind).[18] Both of these, for La Forge,
are species of real causation. For Arnauld, on the other hand, if the mind

15. See, for example, *Examen,* in Arnauld 1775–83, 38:148.

16. In *Principles* 2.40, for example, Descartes speaks quite freely of one body "im-
parting" or "transferring" (the terms he uses are *transfere, dare,* and *transmittere*) motion
to another body. On the other hand, see the letter to Henry More of August 1649, in which
he explicitly denies that motion "transmigrates *[transmigrere]*" from one body to another
(AT 5:405).

17. *Examen,* in Arnauld 1775–83, 38:146.

18. See La Forge 1974, 213. The distinction is, in fact, a medieval one. See, for example,
John Duns Scotus, *Opus oxoniense (Ordinatio)* 1, dist. 2, q. 1, art. 1, pt. 2.

cannot take on motion, then the body cannot be the real cause of *any* effects in the mind. More generally, Arnauld is arguing from the premise that when two substances are as ontologically different as mind and body are, then there is no way for one to causally engage the other and bring about effects therein. The problem, in other words, lies in the dualism.[19]

Arnauld next considers the possibility that God has given to the soul the power to form sensory perceptions for itself and that the soul is occasioned to do so by the appropriate bodily motions. He rejects this solution, however, on the grounds that its conditions clearly are not, and cannot be, fulfilled. In order for the mind to be occasioned by brain motions to produce its own sensations, it would, at the very least, need to perceive or have knowledge of the motions communicated by the nerves to the brain. And we obviously lack such knowledge: "If one should suppose that the soul has such a power to give to itself all the perceptions of sensible objects, it would be impossible for it to produce them so properly [à propos] and with so marvelous a promptitude, since it would not know when it ought to give them to itself, not knowing the corporeal motions that take place in the sense organs that these sense perceptions must follow."[20] Moreover, even if we did know or perceive these motions, we would also need to know the proper sensation to be produced on the occasion of each motion. But given the radical dissimilarity between brain motions and sensory event, as well as our lack of access to some divine lexicon linking the one with the other, we could not know which sensation ought to go with which motions.

Thus, Arnauld concludes, the only remaining alternative is that the direct and immediate cause of our sensations is God, who produces each of them when occasioned to do so by the appropriate motions in the brain. After a warning that one ought not to have recourse to the first cause "without necessity," Arnauld goes on to claim that

> it is not a case of having unnecessary recourse to the first cause when, an effect being certain, one sees that it cannot have any cause in nature. Now it is certain, on the one hand, that our soul never fails, at least ordinarily, to have perceptions of light, colors, sounds, and odors when certain corporeal objects strike our senses; that is to say, as we are informed by anatomy, when they cause motions that travel to the brain. And it is clear, on the other hand, that things being thus, no other natural causes could be

19. Tad Schmaltz has suggested to me that Arnauld's prejudice against the body's acting causally on the mind might also be the result of Augustinian strictures against the less perfect (the body) acting on the more perfect (the mind).

20. *Examen*, in Arnauld 1775–83, 38:147.

assigned to these perceptions that our soul has of sensible objects than either these disturbances that occur in our sense organs, or our soul itself. Now I am able to say that it can be neither the one nor the other, as we have just shown. It only remains for us to understand that it must be that God desired to oblige himself to cause in our soul all the perceptions of sensible qualities every time certain motions occur in the sense organs, according to the laws that he himself has established in nature.[21]

Arnauld's account of body-mind relations with respect to the causal generation of sensations is thus an occasionalist one.[22] Does he think his account is representative of Descartes' views? While there does not seem to be a conclusive answer to this question, I am fairly certain that the answer is no, mainly on the grounds that Arnauld, who usually never misses an opportunity to bolster his case by demonstrating its Cartesian (or Augustinian) pedigree, does not do so here.[23] Moreover, Arnauld is surely aware that in arguing that the mind does not produce its own sensory ideas on the occasion of bodily motions he is thereby arguing against the view Descartes presents in the *Notae in programma*.[24]

21. Ibid., 147–48.

22. See also *Des vraies et des fausses idées*, in Arnauld 1775–83, 38:349. Ndiaye (1991) recognizes that Arnauld's conclusions here are presented in "un vocabulaire parfaitement occasionaliste" and even suggests that for Arnauld corporeal motions are "causes occasionelles" of sensations "comme l'imaginera Malebranche" (309). Yet Ndiaye ultimately grants only that such occasionalist expressions are ambiguous and that the occasionalist tradition in which Arnauld situates himself is more that of Descartes than that of Cordemoy and Malebranche. There is a sense in which Ndiaye's caution is warranted.

23. He does, however, suggest both that his occasionalist account provides a good, solid demonstration of God's existence and that this is the kind of thing that Descartes' philosophy in general does quite well; see *Examen*, in Arnauld 1775–83, 38:150.

24. See AT 8-B:358–59; CSM 1:303–4. Arnauld, however, is also thereby arguing against a view that he himself has elsewhere presented. Thus, there appears to be an incoherence (and it may be only an appearance) in Arnauld's views. While he argues in the *Examen* (1680) that the mind cannot be the efficient cause of its own sensations *(perceptions sensibles, sensations)* on the occasion of bodily motions, he argues in the Port-Royal *Logic* (1662) that the soul is the efficient cause of its ideas *(idées)*, producing them on the occasion of bodily motions: "It may be affirmed, that no idea has its origin in the senses, except by occasion, in that the movements that are made in the brain, which is all our senses can do, give occasion to the soul to form different ideas that it would not have formed without them" (Arnauld 1775–83, 41:133; Arnauld is here self-consciously echoing Descartes' account of innateness from the *Notae*. For a discussion of Descartes' account of the causal origin of ideas, see Nadko, 1994). Has Arnauld changed his account in the intervening twenty years? Is there a significant difference here between sensations and ideas? It might be argued that the account in the *Logic* is to be understood as referring only to clear and distinct ideas, while the occasionalist account of the *Examen* refers to sensations. In *Des vraies et des fausses idées* (1683), Arnauld does suggest that while some "perceptions we

Things start to get a bit tricky when we turn to the context of mind-body relations, particularly in the case of voluntary movements of the body. This is because, unlike his treatment of the body-mind case, Arnauld nowhere argues explicitly and at length for an occasionalist explanation of the mind's working on the body.[25] Nonetheless, there is sufficient and quite good evidence that here, too, Arnauld adopts an occasionalist account.

A convenient passage appears in the third set of objections that Arnauld sends to Descartes:

> You write that our mind has the power to direct the animal spirits into the nerves and in this way to move our members. Elsewhere, you write that there is nothing in our mind of which we are not conscious, either actually or potentially. But the human mind does not seem to be conscious of that power, which directs the animal spirits, since many do not even know whether they have nerves, unless perhaps only nominally, and many fewer whether they have animal spirits, and what animal spirits are. In a word, as far as I can gather from your principles, only that belongs to our mind, whose nature is thought, which belongs to us insofar as we are thinking and aware. But that the animal spirits are directed into the nerves in this or that way does not belong to us insofar as we are thinking and aware. Therefore it does not seem to belong to our mind. In addition, it can scarcely be understood how an incorporeal thing can move a corporeal one.[26]

This passage is convenient not only because in it Arnauld expresses doubts about the mind being the real cause of any bodily motions but also because he does so, in part, for dualistic reasons: "It can scarcely be understood how an incorporeal thing can move a corporeal one." However, there are, for my purposes, two shortcomings with the passage. First, it is rather early, coming from the second of two letters written in June and July 1648, before Arnauld has really established any credentials as a

---

necessarily have from God," others "our soul can give to itself," and he claims that "God gives us the perceptions of light, sounds, and the other sensible qualities . . . on the occasion of what happens in our sense organs or in the constitution of our body" (Arnauld 1775–83, 38:348–49). But I see no evidence in the *Logic* itself for restricting the scope of its account to clear and distinct ideas. In fact, he seems there to be talking about all ideas, clear and distinct as well as sensory.

25. This leads Sleigh to say that "to the best of my knowledge, Arnauld did not reach a conclusion with respect to [mind-body interaction]" (1990, 39). Ndiaye, on the other hand, insists that for Arnauld the mind does exercise real causal influence over the body (1991, 309).

26. Arnauld 1775–83, 38:82–83.

defender of Cartesian philosophy. Second, the skeptical query expressed in the letter is not an argument, much less a statement of occasionalism. The question, then, is whether there is a later and more complete basis for attributing to Arnauld an occasionalist account of mind-body relations.

Yes there is. And the passages in which it is found echo some of the considerations raised in the 1648 letter. In the *Dissertation* (1685), Arnauld argues against what he sees as Malebranche's claims that it is by nature impossible for created intellects to have the power to move bodies. In his argument, Arnauld shows that it is at least *possible* that the human mind should have a real power to move the body. God, he first argues, could (and, in fact, did) give to angels the power to move bodies. Since a spiritual modification is a much more noble thing than a corporeal modification, and since motion is just a modification of a body, there is no reason why an angelic mind, which certainly has the power *(une vertu réelle)* to "move itself" by giving itself new (spiritual) modifications in the form of thoughts and volitions, could not also be given the power to move bodies (cause corporeal modifications). He then generalizes this conclusion to include human minds: if a spiritual creature can modify itself (as human minds can), then there is no reason why it should not be possible for it to have the real power to modify a body.

> A spiritual modification, such as a free act of the will, is a much more noble thing than a corporeal modification, such as the motion or rest of a body. How, then, can one admit that God has given to *all* intelligent natures a real power to form a modification as noble and as excellent as the determination of the will and, at the same time, claim that it is not possible for him to have given to angels a real power to do something as base and contemptible by comparison as change the place of some part of matter? . . . It is not as certain as the author [Malebranche] imagines that God has not given to our soul a real power to determine the course of the spirits toward the parts of the body that we want to move; and it seems that M. Descartes believed it, and that perhaps it is not so easy to prove the contrary.[27]

Yet Arnauld then goes on to argue that, in fact, the human mind does *not* have this power, that it cannot move the body—not because it is, by nature, impossible for a spiritual being to cause a corporeal modification, but because the mind de facto lacks the knowledge required for such causal activity. In particular, the human mind does not know how to move the animal spirits: "Our soul does not know what needs to be done to

27. Ibid., 690; emphasis added.

move our arm by means of the animal spirits. It is properly only this reason . . . that can lead one to believe that it is not our soul that moves our
arm." [28] His reasoning here is similar to that of the 1648 letter, in which
he objects to Descartes' claim that the mind moves the animal spirits on
the grounds that "many do not even know whether they have nerves . . .
and many fewer whether they have animal spirits, and what animal spirits
are." Arnauld does grant that such knowledge, hence such a causal power
over the body, may have been in us before the Fall *(dans l'état d'innocence)* and that it is to be found in the eternally blessed after they are
reunited with their bodies. But in this life the mind does not have the
power to move the body.[29]

Is this an argument for occasionalism? Arnauld does not, in the passages cited, expressly argue for God's active causal role in moving the
body when the mind wills the body to move. But as Arnauld insists in his
argument for body-mind occasionalism, a Cartesian can recognize only
three possible causes of an event: a finite mind, a body, and God. If the
mind cannot move the body, and the body certainly cannot put itself in
motion, then the only possibility is that God moves the body on the occasion of the appropriate mental event (volition). My claim that this is what
Arnauld is leading to here is supported by the fact that it is precisely the
view he presents in *Des vraies et des fausses idées* two years earlier: "We
do not make the least motion, either of a leg or of an arm or of the tongue,
without God himself giving movement to the animal spirits that must
spread into the nerves attached to our muscles in order for that to happen
. . . in that he only executes the general will that he had when he created
us, and . . . it is by our will that the action of God is determined to each
particular effect." [30]

There are two things to be noted about Arnauld's mind-body occasionalism. First, in the argument from the *Dissertation* he does not explicitly
bring in the same kind of dualistic considerations that he uses in arguing
for body-mind occasionalism in the *Examen*. In fact, he tells us that the
only way to demonstrate that the mind cannot act on the body is through
its lack of the requisite knowledge. But I suspect that along with this epistemological argument there remains the mind-body question he raises for
Descartes in 1648: How *could* an incorporeal substance move a corporeal

28. Ibid.
29. Ibid., 690–91. This argument is similar to that offered by Arnold Geulincx, an unequivocal occasionalist. Geulincx argues from the premise that "it is impossible for one to
bring something about if one does not know how it is to be brought about. . . . What one
does not know how to do one cannot do" ([1691] 1892, 2:150).
30. Arnauld 1775–83, 38:285.

one? There is no reason for thinking that Arnauld was ever satisfied by Descartes' response to this objection.[31]

Second, there is conclusive evidence in the argument from the *Dissertation* that Arnauld recognizes that his account of the mind-body relation in voluntary movement is not the one Descartes holds. He admits, that is, that Descartes believes that the mind *is* the real and direct cause of the determination of the motion of the animal spirits and, hence, of the movement of the muscles and limbs. Thus, in both contexts in which Arnauld appeals to an occasionalist solution—mind-body and body-mind relations—he does so in complete awareness that he is departing from the position of his philosophical mentor. Behind both of these occasionalist departures lies his doubt that Cartesian dualism can accommodate real causal interaction between mind and body.

Two objections can be raised against an occasionalist reading of Arnauld. First, if Arnauld is an occasionalist—even a partial occasionalist, as I argue—what then becomes of his celebrated debate with Malebranche? Arnauld expended great energy in attacking Malebranche's occasionalist system, particularly the role Malebranche grants God in the

---

31. For Descartes' response, see his letter to Arnauld of July 29, 1648 (AT 5:222–23; Arnauld 1775–83, 38:86–87). It might be objected that Arnauld's argument in the *Dissertation* constitutes an acknowledgment that an incorporeal substance could conceivably move a corporeal one. Does he not argue that God could give the mind the power to cause a corporeal modification? But I think the conceivability or possibility at work here is an abstract or formal one—divorced from the context of dualism—regarding a hierarchy of powers. If a is more noble than b, and if something S has the power to cause a, then there is no reason per se why S could not also have the power to cause b. But there may be other reasons, extrinsic to the abstract case and when the details are filled in (what a and b are and what kind of thing S is, along with certain metaphysical principles regarding causation), why S cannot conceivably have the power to cause b. Such reasons might be found in Cartesian dualism.

Put another way, Arnauld is not in the *Dissertation* arguing against a view that denies for dualistic reasons that mind can act on body. Rather, he is arguing that if one grants (as Malebranche does) the mind a causal power to produce mental modifications, one cannot then go on to say that it is per se impossible for the mind to have a causal power to produce corporeal modifications—although such a causal power might be ruled out for other reasons (e.g., dualistic ones). If Malebranche grants the mind one causal power, then he should acknowledge the prima facie possibility of its having other, less rigorous or demanding causal powers. What Arnauld is insisting on here is a logical or formal possibility, not a metaphysical possibility. But the mind-body problem is about metaphysical possibility. Thus, it would not be inconsistent for Arnauld to argue both that there is no logical reason why, the mind being capable of causing mental modifications, it cannot cause corporeal modifications, and that there are metaphysical (dualistic) reasons why the mind cannot cause corporal modifications.

causal nexus. On the claim that Arnauld himself holds occasionalist views, his vehement critique of Malebranche appears to become difficult to understand.

There are, in fact, two important aspects of Arnauld's critique of Malebranche's occasionalism, and neither aspect undermines my reading of Arnauld on mind-body relations. On the one hand, what Arnauld is attacking in Malebranche is not divine causal activity in the world per se, but rather Malebranche's claims about the nature of that activity. Malebranche, Arnauld believes, would have God acting only by indeterminate general volitions and never by particular volitions. On this view, God wills only certain general laws and wills that events unfold in complete accordance with those laws, without ever willing any event positively and directly. This mode of activity, Arnauld insists, is unworthy of God and inconsistent with the divine nature.[32] If this is what Malebranche's occasionalism amounts to, Arnauld concludes, then that doctrine is, by Malebranche's own criteria, no better than the pagan systems against which Malebranche is arguing.[33] Indeed, Arnauld even goes so far as to say that Malebranche's conception of occasionalism is idiosyncratic and requires a departure from what is ordinarily understood by *occasional causation*.[34] If God wills something, Arnauld insists, he does so directly and through a particular volition, although that particular volition is itself normally in accordance with a general law and occurs in God in consequence of some other event (as determined by law). If this is the main thrust of Arnauld's attack on Malebranche's occasionalism, then of course there is nothing inconsistent here in Arnauld himself using an occasionalist solution to address certain causal problems.

The other aspect of Arnauld's critique of Malebranche involves Malebranche's claim—a claim essential to his occasionalism—that finite substances, especially human (and angelic) minds, are devoid of any causal efficacy. Arnauld recognizes that Malebranche needs this claim to establish his thesis that "Dieu fait tout." The claim, however, is unacceptable to Arnauld on both philosophical and theological (as well as scriptural) grounds. Putting aside the question of angelic minds (for which, according to Arnauld, Scripture provides sufficient evidence of causal efficacy), Arnauld insists that Malebranche's claim destroys the freedom that all intelligent beings have. This freedom depends upon their being real causes with respect to the determination of their wills toward particular

---

32. This is Arnauld's main concern in book 1 of the *Réflexions*.
33. *Réflexions*, in Arnauld 1775–83, 39:339–48.
34. Ibid., 176.

goods.[35] (Malebranche himself, Arnauld insists, admits as much, and thus there appears to be an inconsistency in his system: Malebranche claims both that God is the sole causal agent and that human minds are free to direct their wills toward particular goods.) Thus, Arnauld rejects an occasionalism that requires God to do everything, including determining our wills. But as in the first case, there is no inconsistency between this aspect of Arnauld's critique of Malebranche's occasionalism and his own use of God's causal activity to bridge the mind-body gap. Note that the occasionalism I am ascribing to Arnauld leaves the mind perfectly free with respect to its volitions. The human mind, he insists, does have a real power to produce its own modifications; it simply cannot act on and move the body.

This brings us to the second objection that might be made against my reading of Arnauld on mind-body relations. According to Arnauld, the mind has real causal efficacy and can form for itself thoughts and volitions. He also appears to allow causal efficacy to bodies, granting that one body can move another body. How, then, can I argue that Arnauld is an occasionalist when he so clearly grants causal powers to finite substances? My reply is that while it is true that an extreme occasionalism such as Malebranche's denies that creatures have any causal efficacy whatsoever, the fact that Arnauld leaves certain causal powers to minds and bodies is irrelevant to the kind of occasionalism I am attributing to him. While the soul can produce mental events and bodies can move other bodies, minds cannot act on (i.e., move) bodies and bodies cannot act on the mind (i.e., cause sensations). This is why God is needed in (and only in) the mind-body context. Arnauld's "occasionalism"—and I use the quotes to indicate that perhaps, with such qualifications, it is improper to call Arnauld an occasionalist in any strict sense of the term—is, thus, an occasionalism with finite causes, minds and bodies that have real causal powers.[36] In the absence of any metaphysical obstacles, those substances can be real causes; because something does, in fact, preclude the action of the one substance upon the other, God steps in and produces the appropriate effect.

Cartesians and Occasionalists

It should be clear from the foregoing that Arnauld's occasionalism is simply an ad hoc solution to a perceived mind-body problem facing the

35. *Dissertation,* in Arnauld 1775–83, 38:686–87. See also *Réflexions* 1.9.
36. As Ndiaye notes, "[L]e vocabulaire qu'utilise Arnauld est occasionaliste. Mais l'inspiration ne l'est pas" (1991, 308–9).

Cartesian dualism he accepts without qualification. In this sense, it fits in quite nicely with the standard textbook mythology about occasionalism. What is surprising, however, is that Arnauld—who is ordinarily not at all thought of as an occasionalist—seems to be the *only* Cartesian who uses occasionalism in this limited way.

Thoroughgoing occasionalists argue for God as the *sole* causal agent, and they do so by showing that finite created substances—minds and bodies—are by nature devoid of any causal efficacy. Malebranche, for example, insists that a causal relationship requires a logically necessary connection and that there is never any such necessary connection between two physical events or two mental events, or between a mental event and a physical event. Thus, no bodily state is the real cause of another bodily state or a mental state, and no mental state is the real cause of any other mental state or a bodily state. He also argues that bodies, as parcels of pure extension, are essentially passive and without active causal power, and that God's role as omnipotent sustaining cause of the world and its contents rules out real secondary causation.[37] Cordemoy's arguments for his occasionalism focus, likewise, on what it is for one being to be the real cause of an effect either in itself or in another being, as well as on what he takes to be the obvious de facto impotence of our minds with respect to our bodies.[38] These and other occasionalists all begin from general considerations regarding the nature of causal relations, the powerlessness of finite beings, the Cartesian metaphysics of matter and motion, or the nature of the relationship between an omnipotent God and his creation. Any specific conclusions about mind-body relations follow from these premises and not from any particular problems inherent in dualism.[39]

Arnauld's limited occasionalism, on the other hand, is not a broad doctrine regarding causal relations generally, but rather a narrow solution to a specific problem generated by dualism. Unlike Malebranche and Cordemoy, Arnauld does not offer any positive philosophical or theological arguments for God's causal role. And he does allow real causal powers in finite substances. In these respects, perhaps it is a technical misnomer to call him an occasionalist. Occasionalism, at a deep level, is a system, a doctrine intended to answer certain metaphysical questions (about being, matter, motion, and even thought). Arnauld's "occasionalism," on the other hand, is neither deep nor systematic—it is a problem-solving device.

37. See Malebranche, *De la recherche de la vérité* 6.2.3 and Eclaircissement 15; also *Entretiens sur la métaphysique,* Dialogue 7.

38. Cordemoy 1968, 135–44.

39. I argue this point in greater detail in Nadler forthcoming. See also Lennon 1974.

But perhaps, as well, the problem lies more in our inadequate conception of seventeenth-century occasionalism. There is not as much uniformity and consistency among occasionalists as we are often led to believe. There are thoroughgoing occasionalists such as Malebranche and Cordemoy, and there are limited or partial occasionalists such as Arnauld, La Forge, and Clauberg. Different thinkers subscribe to this doctrine for different reasons and support it with different kinds of arguments. As far as I can tell, only Arnauld subscribes to it as a solution for (and only for) a mind-body problem.

Let us now return to the question with which this essay began: what kind of Cartesian did the author of the Fourth Objections become? My conclusion, which stands in contrast with the usual picture of Arnauld as an uncritical disciple of Descartes, is that he was, in fact, quite an innovative Cartesian, one who was not above acknowledging and then trying to solve certain problems he perceived in Descartes' metaphysics. The question of mind-body interaction is, for Arnauld, one such problem, and he—alone among Cartesians to recognize it—used an occasionalist solution. While it can be plausibly argued that occasionalism with respect to body-body relations is simply a teasing out of Descartes' own views on matter and motion, I do not see how this can be argued for Arnauld's mind-body occasionalism.[40] With this important matter, then, we have at least one instance in which Arnauld self-consciously adopts a nonorthodox Cartesianism.[41]

40. This is how La Forge, a more orthodox Cartesian than Arnauld, presents his occasionalist account of body-body relations in the *Traité de l'esprit de l'homme* (1666).

41. It makes sense to call Arnauld's account nonorthodox not only because it is a departure from Descartes' views but also because later nonoccasionalist Cartesians (like Régis) accepted real interaction between mind and body.

MARGARET J. OSLER

———————— ᪥10᪥ ————————

DIVINE WILL AND MATHEMATICAL TRUTH: GASSENDI AND DESCARTES ON THE STATUS OF THE ETERNAL TRUTHS

> . . . just as the poets suppose that the Fates were originally established by Jupiter, but that after they were established he bound himself to abide by them, so I do not think that the essences of things, and the mathematical truths which we can know concerning them, are independent of God. Nevertheless I do think that they are immutable and eternal, since the will and decree of God willed and decreed that they should be so.
>
> —René Descartes, Fifth Replies

> . . . the thrice great God is not, as Jupiter of the fable is to the Fates, bound by things created by him, but can, by virtue of his absolute power, destroy anything that he has established. . . .
>
> —Pierre Gassendi, *Disquisitio metaphysica*

Pierre Gassendi (1592–1655) and René Descartes (1596–1650) were two of the major advocates of the mechanical philosophy during the first half of the seventeenth century. Both made seminal contributions to the view that all the phenomena of the physical world should be explained by reference to matter and motion. Agreeing on this fundamental tenet of the mechanical philosophy as well as on their rejection of the Aristotelian and occult philosophies of the day, they disagreed about virtually everything else: the nature of matter, the epistemological status of scientific knowledge, and the particular mechanical explanations of individual phenomena. Gassendi, following the ancient models of Epicurus and Lucretius, maintained that indivisible atoms and the void are the ultimate components of nature. Atoms possess magnitude, figure, and heaviness, properties that cannot be fully known by reason alone. He advocated an empiricist theory of scientific knowledge and a nominalist ontology, claiming that only individuals exist and that universals have no independent, extramental reality. Descartes, by contrast, claimed that the universe is a plenum and that the matter filling it is infinitely divisible. According to him, matter is identical with geometrical space, and its only property is extension, an attribute that can be understood rationally, without any appeal to observation or experience. Although his theory of scientific knowledge requires appeal to empirical methods, he claimed that the first principles of natural philosophy can be known a priori. In this essay I will argue that the differences between Descartes and Gassendi—focusing here on their philosophical interpretations of mathematics—can be understood in light of their respective understandings of God's relationship to the creation. Their theological views have long roots, extending back into the Middle Ages, and their differences continue to reflect medieval discourse about the nature of divine power.

Increasing contact between Greek philosophy and Christian theology in the twelfth and thirteenth centuries led to the formulation of two different understandings of the relationship between divine power and the natural order God created. These theological positions are most commonly called voluntarism and intellectualism, and their meaning is usually explained in terms of the relationships between two of God's attributes, his intellect and his will.[1] On this account, voluntarists stress the primacy of God's will over his intellect, and intellectualists emphasize God's intellect over his will. Because theologians from both camps agree that God has both intellect and will, the differences between them are subtle.

Rather than defining these theological positions in terms of the relationship between the divine attributes, I find it useful to think of the differences between them by considering God's power in relation to the creation. Medieval theologians used the terms *potentia Dei absoluta* and *potentia Dei ordinata*—God's absolute power and God's ordained power—to capture that relationship.[2] These terms do not refer to two powers that God possesses, but rather to two ways of understanding divine power. God's absolute power refers to what is theoretically possible for him to do providing it does not involve a logical contradiction. It refers to divine power apart from any particular acts God chooses to perform. In establishing the present world order, God exercised his absolute power. In so doing, he did not exhaust the creative possibilities open to him. In this sense his power is unbounded.[3] *Potentia ordinata* refers to divine power with respect to what God has actually chosen to do in establishing the present order.[4] God has created a world filled with various kinds of

1. See Foster 1936; and Taylor 1967.
2. On the history of this distinction, see Desharnais 1966.
3. Courtenay 1985, 243.
4. Ibid., 247.

things, and he has created the laws governing the relationships among them. His *potentia ordinata* is his power vis-à-vis his creation, his governance of the world in accordance with the way he created it, and his ability to intervene in the created order.

In these terms, the difference between intellectualism and voluntarism can be expressed by asking how binding the created order is on God's present and future acts. Intellectualists were prepared to accept some necessity in the creation, while voluntarists regarded the present order as utterly contingent. Viewed in this way, there were many different kinds of intellectualists. Some, such as Peter Abelard, took an almost Platonic position in claiming that a rational order exists independent of God.[5] Platonic forms and Scholastic uncreated essences, such as those discussed by the sixteenth-century Jesuit Aristotelian Francisco Suárez, are both examples of such entities existing independent of God. Other intellectualists, such as Thomas Aquinas, took a more moderate position in allowing that while nothing exists that God did not create freely, nevertheless some created beings contain necessary relations that render a priori demonstrative knowledge of at least some aspects of the creation possible.[6]

In contrast to intellectualists, voluntarists insisted on God's omnipotence and his absolute freedom of will: nothing exists independent of him, and nothing that he created can bind or impede him. Certainly a rational order independent of God would reduce the scope of his absolute power, thereby limiting his freedom of action in the world. Even the necessity of laws he created freely would restrict the exercise of his power over the creation. Arguing in this vein, Peter Damian contended that God in his omnipotence created everything, even the laws of logic, which he can change if he so wills.[7] Apart from Damian's extreme position, most voluntarists regarded the law of noncontradiction as the only exception to God's absolute freedom in order to ward off such potentially devastating absurdities as God's willing himself to cease existing.

The voluntarists' emphasis on the contingency of the creation had important epistemological implications. Without necessary relations, it is not possible to attain a priori, demonstrative knowledge of the creation.

5. Oakley 1984, 45.

6. "A rationalist theology involves both a rationalist philosophy of nature and a rationalist theory of knowledge of nature. If God made the world according to reason, the world must embody the ideas of his reason; and our reason, in disclosing to us God's ideas, will at the same time reveal to us the essential nature of the created world" (Foster 1936, 10). On the varieties of intellectualism, see Wells 1961.

7. Courtenay 1985, 244–45; Taylor 1967, 271. See also Oakley 1984, 42–44.

Such knowledge is not possible because any guarantee that the contents of the human mind must correspond to the world would involve the existence of some kind of necessary relations, which are unacceptable to voluntarists because of their emphasis on God's absolute power. Any regularities that are observed are simply that, observed regularities. The natural order may be regular, but it is also completely contingent. Since God's freedom suffers no restraint, he can alter the observed regularities at will, a possibility to which miracles attest. Consequently, any knowledge we can have of the creation is fallible and has only observation as its source, for there is no guarantee that the course of nature will be constant or must correspond to the limited capacities of human understanding. It follows—for voluntarists—that all human knowledge of nature must be empirical. Moreover, nothing about the creation can ever be known with certainty: God's absolute power renders human knowledge fallible and, in this sense, only probable. Either nominalism or conceptualism—in any case, a metaphysics that denies independent ontological status to universals—is also the regular concomitant of voluntarist theology. Only particular individuals exist in the world, and no necessary relations connect them. Consequently, any relations we attribute to groups of them—such as similarity—are the product of our minds. They do not exist independently in the world.

The differences between Gassendi and Descartes on the status of mathematical truths is an expression of their differences about God's relationship to the creation.[8] Gassendi was a voluntarist, interpreting God's omnipotence to mean that he is free from any necessity or limits. "[T]here is nothing in the universe that God cannot destroy, nothing that he cannot produce; nothing that he cannot change, even into its opposite qualities."[9] In other words, God's absolute power is in no way constrained by the creation, which contains no necessary relations that might limit God's power or freedom. Even the laws of nature lack necessity, since he can negate them like everything else he created. "[H]e is free from the laws of nature, which he constituted by his own free will."[10] Indeed, God can do anything short of violating the law of noncontradiction. God was totally free in choosing to create the world: he could have abstained from creating it just as freely as he chose to create it. Moreover, God could have created an entirely different natural order, if it had pleased him to do so.[11]

8. For a more thorough discussion of all of these issues, see Osler 1994.
9. Gassendi, *Syntagma philosophicum*, in Gassendi [1658] 1964, 1:308.
10. Ibid., 381; see also 234.
11. Ibid., 1:309, 318, 2:851; see also 1:307.

That is to say, God could have created black snow, and he could have created a cold fiery substance, just as he can raise the dead and heal the crippled. The order that God created is utterly contingent on his absolute power. Even though he has chosen to create the universe—the one containing white snow and hot fire—he has created nothing in it that he cannot change at will. There are no necessary connections linking fire and heat or whiteness and snow. God could not have made white snow black or hot fire cold, for such combinations of attributes are contradictory; but he could have created substances with properties very different from the ones that presently exist. And by implication, he is free to do so at any time. The inability to create contradictories lies not in a restriction of God's power but in a repugnance in the things themselves. The fact that we cannot understand God's power does not mean that God can do nothing beyond our understanding. It would be presumptuous for us to limit God to the scale of our intellect.[12] "Even the fact that God responds to our prayers in no way limits his freedom: when God is beseeched, he acts because he determines to; and when it is said he is bent by prayers, the mode of speaking is vulgar . . . for we only give him prayers that he constituted himself. Therefore God, as he is the highest good, is the most free; and he is not bound, as he can do whatever he knows, whatever he wishes."[13]

With the stipulation that nothing God creates can impede his absolute power, God makes use of second causes to carry out the ordinary course of nature: "[I]t is his general providence that establishes the course of nature and permits it to be served continuously. From which it follows, as when either lightning or other wonderful effects are observed, God is not on that account suddenly summoned, as if he alone were its cause and nothing natural had intervened. . . . [A]side from him, particular [causes] are required that are . . . not thought to be uncreated, but are believed to be hidden from our skill and understanding."[14] Second causes, as part of the created order, do not restrict God's freedom, because God can dispense with second causes altogether if he chooses. "Indeed, to the extent that an artisan has ingenuity, to that extent his work requires less matter and services; thus the divine artisan will be the most skilled, since he requires neither matter nor ministers nor organs nor equipment."[15] Here Gassendi sounds much like Ockham, who said, "Whatever God can produce by means of secondary causes, He can directly produce and pre-

12. Ibid., 1:308, 309; see also 381. 14. Ibid., 326.
13. Ibid., 309. 15. Ibid., 317.

serve without them." [16] According to Gassendi, God is thus free, even from the things he has created, "since he neither is confined by anything nor imposes any laws on himself that he cannot violate if he pleases. . . . Therefore God . . . is the most free; and he is not bound, as he can do whatever . . . he wishes." [17] The natural order, which God created by his absolute power, remains utterly contingent on his will.

In contrast to Gassendi's voluntarism, Descartes drew on an intellectualist theological tradition that he absorbed from the Jesuit Aristotelianism to which he was exposed when he was a student at La Flèche.[18] Descartes expressed his views on God's relationship to the creation most fully in several letters he wrote to Mersenne in April and May 1630, in which he discussed the status of the so-called eternal truths. In these letters he espoused an intellectualism about the ordained, not the absolute, power of God.

Descartes began by asserting that the eternal truths, like everything else in the world, have been created by God. "The mathematical truths which you call eternal have been laid down by God and depend on Him entirely no less than the rest of his creatures." Like Gassendi's laws of nature, they can be understood as the products of God's absolute power. "Please do not hesitate to assert and proclaim everywhere that it is God who has laid down these laws in nature just as a king lays down laws in his kingdom." [19] To think of the eternal truths as existing independent of God would be a blasphemous affront to divine omnipotence. Unlike Plato's Demiurge or the pagan gods, God is not limited by anything else in the world.[20] "Indeed to say that these truths are independent of God is to talk of him as if he were Jupiter or Saturn or to subject him to the Styx and the Fates." [21] From the standpoint of his absolute power, God is subject to nothing. His freedom is complete.

God not only created the eternal truths, but he also created our minds in such a way that we possess an innate capacity to understand them. Here Descartes introduced a major element of necessity into the world, the connection between the eternal truths and our minds. From rather traditional talk about the powers of God, he produced the metaphysical groundwork for what became the basis of his revolution in philosophy: "There is no single one [of the eternal truths] that we cannot understand

16. William of Ockham, *Ordinatio* Q. i, N sqq.; William of Ockham 1957, 25.
17. Gassendi, *Syntagma philosophicum,* in Gassendi [1658] 1964, 1:309.
18. On the curriculum at La Flèche, see Garber 1992, 5–9; and Dear 1988, 12–15.
19. AT 1:145; trans. Kenny, in Descartes 1970, 11.
20. See Cornford [1937] 1957, 38.
21. AT 1:145; Descartes 1970, 11.

if our mind turns to consider it. They are all *inborn in our minds* just as a king would imprint his laws on the hearts of all his subjects if he had enough power to do so."[22] If the eternal truths are somehow inborn in our minds, how can we be assured that God will not change them, that we will not be deceived in believing them to be true? "It will be said that if God had established these truths he could change them as a king changes his laws. To this the answer is: 'Yes he can, if his will can change.' 'But I understand them to be eternal and unchangeable.'—'I make the same judgment about God.'—'But his will is free.'—'Yes, but his power is incomprehensible.'"[23]

By his absolute power, God freely created the eternal truths, just as he freely created the other creatures. In this act of creation, God's power and freedom were absolute, unconstrained by anything else in the universe. Quite simply, nothing in the universe exists independent of God, even the laws of mathematics. "You ask also what necessitated God to create these truths; and I reply that just as he was free not to create the world, so he was no less free to make it untrue that all the lines drawn from the centre of a circle to its circumference are equal."[24] Thus, the eternal truths are true because God knows them. It is not the case that he knows them because they are true, for in that case both the truths and an absolute standard of truth would exist independent of God's creative act. "As for the eternal truths, I say once more that *they are true or possible only because God knows them as true or possible. They are not known as true by God in any way which would imply that they are true independently of him.* If men really understood the sense of their words they could never say without blasphemy that the truth of anything is prior to the knowledge which God has of it."[25] Once these truths have been created, however, they are eternal, even though God's will remains free, for God's will is immutable. "In God willing and knowing are a single thing in such a way that *by the very fact of willing something he knows it and it is only for this reason that such a thing is true.* So we must not say that if *God did not exist nonetheless these truths would be true;* for the existence of God is the first and most eternal of all possible truths and the one from which alone all others derive."[26] Note that the existence of God is an

22. Ibid.
23. AT 1:145–46; Descartes 1970, 11–12.
24. Descartes to Mersenne, May 27, 1630, in AT 1:152; Descartes 1970, 15.
25. Descartes to Mersenne, May 6, 1630, in AT 1:149; Descartes 1970, 13.
26. Descartes to Mersenne, May 6, 1630, in AT 1:149–50; Descartes 1970, 13–14. Harry Frankfurt points out that by uniting God's will and intellect, Descartes rejected the Scholas-

uncreated eternal truth, but obviously not one that in any way limits divine freedom.[27]

Descartes believed that the immutability of God's will is entailed by the unity of his will and his understanding. If an act of willing (say, to create the eternal truths) is simultaneously an act of understanding (of these same truths), then a change of his will would entail a change of his understanding. But any change in the divine understanding would entail some imperfection in God: his understanding would at some time—before or after the change—have been mistaken or incomplete. That is to say, something he knew at one time he would not know at some other time. It would then not be possible to say that he has perfect knowledge. Such imperfection would be incompatible with divine perfection. Consequently, it is not possible to assert without contradiction that God's will can change. The immutability of God's understanding and the unity of his will and intellect jointly entail the immutability of his will.[28] Furthermore, divine immutability provides Descartes justification for the necessity of the eternal truths that God created freely.[29]

To recapitulate, according to Descartes, God freely willed to create the eternal truths and the laws of nature. This act of creation—a product of God's absolute power—established the order of nature, which, because of the immutability of his will and the unity of his intellect and will, he maintains and conserves in the manner in which he first created it. The ensuing necessity of these truths and laws is a necessity relative to this natural order and to our understanding of it. Like Aquinas' necessity of

---

tic view that possible essences are "dependent on God's understanding but independent of His will." By identifying the divine will and intellect, Descartes was led to regard freedom as indifference. "Since there *are* no truths prior to God's creation of them, His creative will cannot be determined or even moved by any considerations of value or rationality." See Frankfurt 1977, 40–41. Wells (1982) criticizes Frankfurt's argument that the eternal truths are "mind dependent" on the grounds that these uncreated truths have nothing to do with the human mind. See also Descartes, Sixth Replies, in AT 9-A:232–34.

27. Wells 1982, 185–99.

28. On the interconnection between God's will and understanding, see Gilson 1913, 34–48.

29. In making this argument, Descartes shows a striking similarity to Thomas Aquinas. Aquinas wrote that "everything eternal is necessary. Now, that God should will some effect to be is eternal, . . . [and] is therefore necessary. But it is not necessary considered absolutely, because the will of God does not have a necessary relation to this willed object. Therefore it is necessary by supposition" (1957, 1:263). For Aquinas, God is free in his choice to will something into eternal existence; but once having been created, that thing is both eternal and necessary. This similarity between Aquinas and Descartes underscores the resonance of the dialectic of God's absolute and ordained power in Descartes' thinking.

supposition, it can thus be understood in terms of God's ordained power. "And even if God has willed that some truths should be necessary, this does not mean that he willed them necessarily; for it is one thing to will that they be necessary, and quite another to will them necessarily, or be necessitated to will them." [30] According to Descartes, the eternal truths are necessary, even though God created them freely and their existence depends entirely on him. Our ability to know them a priori, moreover, entails the existence of a necessary relation between at least some of the ideas in our mind and their external referents.

Descartes' position is an intellectualism with regard to the ordained, not the absolute, power of God. The differences between this position and the voluntarism of Ockham, the nominalists, and Gassendi are subtle. Although both groups denied the existence of anything that God did not create, they differed as to the status of the order that he did. In essence, the voluntarists equated God's absolute and ordained power so that his ordained power was absorbed into his absolute power, leaving nothing beyond his direct control. In contrast to the voluntarists' insistence on the utter contingency of the creation, Descartes, like Aquinas and his followers, believed in the existence of at least some elements of necessity in the created world. The relative necessity of Descartes' eternal truths closely resembles Aquinas' idea of necessity by supposition.[31] This necessity resides in the created order and delineates an area where God's absolute and ordained power do not overlap. It is a mode that no voluntarist would have countenanced.

God's creation of the eternal truths provided the metaphysical and epistemological foundations for Descartes' a priori, demonstrative approach to the first principles of philosophy.[32] Descartes believes that in creating this world God established certain fundamental principles that are necessarily true and the truth of which we can prove a priori. Descartes' claim to have a priori, certain knowledge constituted his response to the skeptical challenge. Rather than working within a general acceptance of the skeptical arguments, as Gassendi did, Descartes wanted to defeat skepti-

---

30. Descartes to Mesland, May 2, 1644, in AT 4:118–19; Descartes 1970, 151. Again, note the striking parallel with Thomas Aquinas: "Now, the fact that God is said to have produced things voluntarily, and not of necessity, does not preclude His having willed certain things to be which are of necessity and others which are contingently, so that there may be an ordered diversity in things. Therefore, nothing prevents certain things that are produced by the divine will from being necessary" (1957, 2:86).

31. See, for example, Thomas Aquinas 1957, 2:85.

32. Marion argues convincingly that the letters of 1630 on the status of the eternal truths determine the point of departure for the *Meditations*. See Marion 1981, chap. 14.

cism by finding an indubitable kernel of certainty from which to construct his deductive system. He found the ultimate justification for this certainty in his doctrine of the eternal truths.

Descartes' arguments in the *Meditations*—systematic doubt, the *cogito,* the proofs of the existence of God, the soul, and matter—are well known and need not be reiterated here. What may be less well known is the connection between these arguments and his theory of the creation of the eternal truths. Having employed systematic doubt to eliminate any dubitable claims, Descartes discovered the certainty he sought in the proposition *cogito ergo sum.* In order to proceed from the *cogito* to the existence of the physical world, he needed a criterion by which to judge the reliability of his knowledge. He found that criterion in the principle that "everything that we clearly and distinctly understand is true in a way which corresponds exactly to our understanding of it."[33] Lying at the heart of both the *Meditations* and the *Principles of Philosophy,* that principle follows from the fact that if what we perceive clearly and distinctly turned out to be false, God would be a deceiver, an attribute incompatible with divine perfection.[34] The clear-and-distinct principle can thus be understood as one of the eternal truths that God implanted in our minds. The truth of this principle follows from the divine nature, and it functions as the necessary connection between the contents of our minds and the natural order God created by his absolute power. The criterion of clear and distinct ideas served as the main underpinning for Descartes' claim to have a priori knowledge of the nature of matter as extension, its infinite divisibility, and the doctrine of primary and secondary qualities, the fundamental components of his mechanical philosophy of nature.

Despite the fact that Gassendi and Descartes shared a strong anti-Aristotelianism, a consuming interest in the new science, and a dedication to the establishment of a new, mechanical philosophy of nature, their arguments following the publication of Descartes' *Meditations* in the ensuing *Objections* and *Replies* and amplified in Gassendi's *Disquisitio metaphysica* (1644) were heated, to say the least. This controversy signaled a major rift between the two founding fathers of the mechanical philosophy. What was really at stake between them?

Among other things, Gassendi and Descartes disagreed about the status of universals and mathematical truths, and here their argument circles back to their implicit theological differences. Descartes believed that eter-

33. Descartes, *Meditations,* in AT 7:13; CSM 2:9.
34. Descartes, *Principles of Philosophy,* in AT 8-A:16 and 9-B:38 (see also AT 8-A:20–23 and 9-B:42–50). Note that this solution is not open to Ockham because of his conception of divine liberty. See Adams 1970, 397.

nal and immutable natures exist, a view he affirmed in the Fifth Meditation:

> When . . . I imagine a triangle, even if perhaps no such figure exists, or has ever existed, anywhere outside my thought, there is still a determinate nature, or essence, or form of the triangle which is immutable and eternal, and not invented by me or dependent on my mind. This is clear from the fact that various properties can be demonstrated of the triangle, for example, that its three angles equal two right angles, that its greatest side subtends its greatest angle, and the like; and since these properties are ones which I now clearly recognize whether I want to or not, even if I never thought of them at all when I previously imagined the triangle, it follows that they cannot have been invented by me.[35]

His ability to demonstrate various properties of the triangle led him to conclude that the triangle has an "immutable and eternal essence," independent of his own mind. The essence of the triangle thus has the status of an eternal truth, created freely by God, but created to be necessarily true. The necessity embedded in the nature of the triangle provided grounds for Descartes' claim to have demonstrative, a priori knowledge of its properties.[36] Descartes' conclusion depends upon an important assumption that Gassendi denied: namely, that the idea of the triangle I have initially is the very same idea I have after I have demonstrated some of its properties. Gassendi held that our idea of a triangle changes as we uncover new properties of the triangle. He was thus in a position to deny that such demonstrations show that we grasp a fixed and immutable essence of the triangle before we demonstrate anything about it.

Gassendi, the voluntarist, found it "hard to agree that there exists some *immutable and eternal nature* other than [that of] omnipotent God."[37] Descartes replies that God in fact created them that way: "I do not think that the essences of things and mathematical truths can be known to be independent of God, but I also think that because God wished it thus, and has thus disposed them, they are immutable and eternal."[38] Here Descartes returned to his theory of the created eternal truths.

Gassendi replied to Descartes' theory of the eternal truths by focusing

35. AT 7:64; CSM 2:44–45.

36. The equivocation here between the necessary relation between the essence and properties of a triangle, on the one hand, and the necessity of propositions about the triangle, on the other, is Descartes'.

37. Gassendi [1658] 1964, 3:374; Gassendi 1962, 468–69.

38. Gassendi [1658] 1964, 3:375; Gassendi 1962, 472–73.

on its connection to God's role as creator, arguing that Descartes had produced an unresolvable dilemma:

> [W]hen you say that "God wished that these natural realities be immutable and eternal and that he disposed them thus," either these natures in question existed by themselves and were not created, and God willed them and disposed them simply as a worker disposes to his will in building with the stones he finds already made and not created; or they were created by God, and God was their author in the sense that he wished and disposed them by a creative act.[39]

One horn of the dilemma places unacceptable limitations on God's absolute power by giving these eternal and immutable natures an existence independent of him. "In the first case, these natures exist by themselves, independent of God; and it is uselessly that God wishes them thus, since they were such by themselves." The other horn of the dilemma allows for divine creation of these essences but thereby implies that they are not eternal: "In the second case, they would not be eternal since in order to be created they must not have been and have thus in a given moment commenced to be, which is, for an eternal thing, contradictory."[40] Gassendi's analysis of the problem thus led directly to the question of God's absolute power and its relationship to the world. Gassendi correctly perceived that Descartes was an intellectualist, although in this case he portrayed Descartes as a more extreme intellectualist than he actually was. Descartes' theory of the creation of the eternal truths had given those truths a necessity relative to God's having created them by his absolute power. Gassendi interpreted Descartes' reasoning about the essence of the triangle to mean that essences have an existence prior to and independent of God's absolute power. This Platonic position is not the one that Descartes had enunciated in his other writings, although it is an easy conclusion to draw from his statement that the triangle's nature is eternal. In either case, the two philosophers disagreed on questions about God's power and the presence of necessity in the world.

In objecting to Descartes' assertion that mathematical truths are eternal and immutable, Gassendi reiterated his nominalism and empiricism, to wit, his view that mathematical concepts and, more generally, statements about universals are abstractions drawn by the human mind from the sensory experience of particulars. "The triangle that you have in your mind is a kind of rule by means of which you examine if a thing merits

39. Gassendi [1658] 1964, 3:377; Gassendi 1962, 480–81.
40. Ibid.

the name of a triangle, but it is not necessary to say for that that this triangle is something with a real and true nature outside of the understanding: only that after having seen material triangles, it has formed this nature and rendered it common."[41]

Descartes had no patience with Gassendi's view. He replied that it is impossible to form ideas of mathematical figures from experience, because there are no truly straight lines or dimensionless points in the material world:

> But because the idea of a true triangle is nevertheless in us and it can be more easily conceived by our mind than the more complex figure of the drawn triangle, it follows that the complex figure, having been seen, is not the same as the former, but rather the true triangle that we have grasped by thought. . . . It is certain that we could not recognize the geometrical triangle by the manner in which it is traced on paper unless our mind had received the idea in some other way.[42]

Since the material representation of the triangle is inadequate for teaching us its true nature, we must know this nature because God implanted it in our minds as one of the created eternal truths.

The debate boiled down to questions about the relative importance of reason and observation and the extent to which human understanding can penetrate the essences of things. Gassendi unequivocally rejected Descartes' realism and rationalism: "[B]esides the thrice-great God, we do not know a single thing that was not created by him . . . really singular."[43] And he perceived Descartes' position as dangerous to the true faith because its essentialism placed unacceptable limits on God's freedom and power. Quoting Giovanni Pico della Mirandola, Gassendi stated, "Nothing is more dangerous for a theologian than to know the *Elements* of Euclid."[44]

The issues that separated Gassendi and Descartes stemmed directly from their divergent understandings of God's relationship to the creation. I argue that Gassendi's voluntarism and Descartes' intellectualism were embedded in their respective philosophical views about the nature of the world and the status of our knowledge. Not surprisingly, their confrontation following the publication of the *Meditations* found them at loggerheads on these very issues. Their confrontation becomes more fully intelli-

41. Gassendi [1658] 1964, 3:375; Gassendi 1962, 470–71.
42. Gassendi [1658] 1964, 3:376; Gassendi 1962, 474–77.
43. Gassendi [1658] 1964, 3:377; Gassendi 1962, 482–83.
44. Gassendi [1658] 1964, 3:384; Gassendi 1962, 514–15.

gible when it is placed in the context of their theological differences. These differences had further implications when Gassendi and Descartes articulated their versions of the mechanical philosophy. Their respective accounts of the properties of matter and the status of mechanical explanations are closely related to their theories of knowledge and existence.

THOMAS M. LENNON

᎓ 11 ᏊᏫ

PANDORA; OR, ESSENCE AND REFERENCE:
GASSENDI'S NOMINALIST OBJECTION
AND DESCARTES' REALIST REPLY

Do not argue with the people of the Book
unless in a fair way,
. . . and say to them:
"We believe what has been sent down to us,
and we believe what has been sent down to you.
Our God and your God is one,
and to Him we submit."

—Al-Qur'an 29.46

Call him Allah [Power] or call Him Ar-Rahmant [Mercy]
whatever the name you call Him by,
all His names are beautiful.

—Al-Qur'an 17.110

The objections of Gassendi, which Mersenne sent to Descartes on May 19, 1641, produced a significant change in Descartes' attitude toward him. Previously Descartes respected Gassendi as, if not an authority, then at least someone whose ideas concerning optics, astronomy, and other matters were to be regarded seriously.[1] On June 23, 1641, however, he returned to Mersenne Gassendi's objections along with the advice *(entre nous)* to have them printed without showing them to Gassendi, who when he saw Descartes' reply would want them "suppressed." Descartes meanwhile was loath to see the time he had taken to reply wasted; nor did he want anyone to think that it was he who, unable to answer the objections, had them suppressed. He concluded the letter, "[Y]ou will see that I have done all I could to treat Gassendi honorably and gently; but he gave me so many occasions to despise him and to show he has no

1. See letters to Mersenne, Dec. 18, 1629, Jan. 1630, March 4, 1630, May 6, 1630 in AT 1:97, 112–13, 127, 148; see also Dec. 1638, in AT 2:464–65.

common sense and can in no way reason, that I would have done too much less than my duty had I said less, and I assure you I could have said much more." [2]

A month later Descartes again wrote to Mersenne, saying that Gassendi had no grounds for complaint at his treatment, for he gave Gassendi only equal in kind despite what he had always heard, namely, that the first blow is worth two and that thus to be really equal the reply should have been doubled. "But perhaps he was affected by my replies because he recognized their truth, while I was not for an entirely different reason; if so, it is not my fault." [3] Two years later Descartes could still muster respect for Gassendi's empirical astronomy, but by then he could tell just from the index of Gassendi's letters that they contained nothing he needed to read. [4] The literature has tended to fault Gassendi for failing really to enter into Descartes' views; the converse seems no less true. For the most part Descartes treated his would-be adversary as beneath contempt. Discussing why criticism of his work was no burden, Descartes wrote that Gassendi's criticisms did not displease him as much as he was pleased by Denis Mesland's judgment that there was nothing in them not easily answered. [5]

As for Gassendi's attitudes, there are two versions of the story. One is that he was nothing but livid with Descartes and that he was no less acrimonious and petulant. But this version is based on the testimony of Jean-Baptiste Morin, an eccentric to say the least, who on other grounds was concerned to besmirch the reputation of Gassendi. Thus, for example, Descartes' calling Gassendi "Flesh" in his *Replies* would be understood only as justified retaliation for having been referred to as "Mind" in the *Objections*. More credible is the account of Gassendi's student François Bernier, who portrays a rather more detached, long-suffering response from his teacher. In any case, the personal differences between the great antagonists were finally repaired in 1648 following a dinner arranged for them by the abbé d'Estrées. As it happened, Gassendi was unable to attend because of illness; but after

2. AT 3:388–89. In the later appendix to the Fifth Replies, Descartes says that Gassendi's objections "did not seem to [him] to be the most important, and [that] they were extremely long; but nonetheless [he] agreed to have them published in their appropriate place out of courtesy to their author. [He] even allowed [Gassendi] to see the proofs, to prevent anything being printed of which he did not approve" (AT 9A:198; CSM 2:268). Either there is more to the story, or Descartes was plainly disingenuous.

3. To Mersenne, July 22, 1641, in AT 3:416.

4. To Colvius, April 23, 1643, in AT 3:646; to Mersenne, Feb. 23, 1643, in AT 3:633.

5. To Noel, Dec. 14, 1646, in AT 4:585–86.

the dinner, the assembled party visited Gassendi, who was embraced by Descartes.[6]

Gassendi's objections may well be regarded as unique.[7] Among other objectors, Arnauld, who really asks only for clarifications, and the Scholastic theologians share important metaphysical presuppositions with Descartes; and Hobbes does not engage Descartes so much as mechanically juxtapose his own views to those of Descartes, with Descartes replying in kind. In the case of Gassendi, however, we find an elaborated and systematic metaphysical confrontation. The length of his exchanges with Descartes thus reflects their relative importance. The Fifth Objections (1641) are more than twice as long as any other set, and if Descartes thought them "not the most important," he nonetheless replied to them at greatest length.[8] Within a year Gassendi had responded with his Rebuttals (*Instantiae:* literally, "follow-ups"), which with the Fifth Objections and Replies were published in 1644 under the general title of *Disquisitio metaphysica,* totaling some 150 pages *in folio* of the *Opera omnia.* Finally, there is Descartes' appendix to the Fifth Objections and Replies, first published in 1647 with the first French edition. It consists of a fifteen-page letter to Claude Clerselier, preceded by an author's note in which Descartes says that upon rereading Gassendi's objections and rebuttals, he was unable "to discover a single objection which those who have some slight understanding of my *Meditations* will not . . . be able to answer quite easily without any help from me."[9] In the letter itself, Descartes responds to objections culled from the Rebuttals by Clerselier's friends.

A handle on this mass of material is provided by a single text from Descartes' initial reply to Gassendi. The text concerns the indivisibility of essences, and it permits a systematic confrontation of two philosophies that are radically opposed at the deepest level, on a range of issues, including the nature of ideas and knowledge, of essences, reference, and explanation. Although it is here possible at best to give only an indication of the broader systematic significance of this text for Descartes and Gassendi, it will yet be possible to explore it somewhat in Locke, Malebranche, and others.

---

6. The episode is recounted in Bougerel 1737, 306–8. See Howard Jones, who gives more details on these personal relations (1981, 66–69) and who also gives the substance of Gassendi's criticisms of the *Meditations* (135–88).

7. I have benefited greatly from the work of O. R. Bloch, to whom my debt will be apparent. Bloch first dealt with the topic in "Gassendi critique de Descartes" (1966); he later elaborated many of the themes in *La philosophie de Gassendi* (1971).

8. AT 9-A:198; CSM 2:268.

9. AT 9-A:199; CSM 2:269.

The Key Text: The Indivisibility of Essence

Having argued the existence of God on the basis of his own existence and his own idea of God, Descartes in the Third Meditation then examines how *(qua ratione)* he received this idea of God from him. He did not derive this idea *(neque . . . hausi)* from the senses, for it did not come to him unexpectedly, as the ideas of sensible things ordinarily do. Neither is the idea of his own making *(nec etiam a me efficta est)*, and therefore it must be innate. The argument that it is not a factitious idea is that he is unable to take away anything from it or to add anything to it *(nam nihil ab illa detrahere, nihil illi superaddere plane possum)*, as he would be able to do if it were factitious.[10]

Gassendi picked up on this argument for innateness and objected that the idea of God could be accounted for as partly derived from the senses and partly of Descartes' own making. "When you say that you cannot add anything to it or take anything away, remember that when you first acquired it, it was not as perfect as it is now." Indeed, our knowledge of the perfections of God that we derive from the perfections of created things is in principle always open to further improvement. At no time, according to Gassendi, do we possess a perfect idea of God.[11]

The brevity of Descartes' reply belies its importance.

> When you attack my statement that nothing can be added to or taken away from the idea of God, it seems that you have paid no attention to the common philosophical maxim that the essences of things are indivisible. An idea represents the essence of a thing, and if anything is added to or taken away from the essence, then the idea automatically becomes the idea of something else. This is how the ideas of Pandora and of all false Gods are formed by those who do not have a correct conception of the true God.[12]

To explain how it is that we nonetheless are able continually to detect additional perfections in God, Descartes proposes not an augmentation, but an explication or clarification of the idea of God. If the idea is a true one, then all the perfections we gradually notice in it are already contained in it and all we do is to make them distinct. "Similarly, the idea of a triangle is not augmented when we notice various properties in the tri-

10. AT 7:51; CSM 2:35.

11. AT 7:305; CSM 2:212. Not irrelevantly, Locke gives a similar account of our idea of God. The mind "enlarges upon" ideas of existence and duration, knowledge and power, and so on, that are ultimately derived from experience, making them "boundless," that is, (potentially) infinite. See *Essay* 2.23.33–35, in Locke 1975, 314–15.

12. AT 7:371; CSM 2:255–56.

angle of which we were previously ignorant." Our first realization of the Pythagorean theorem, for example, does not give us a new idea of the right-angled triangle, but a more distinct idea of it. "The idea of God is not gradually formed by us when we amplify the perfections of his creatures; it is formed all at once and in its entirety as soon as our mind reaches an infinite being which is incapable of any amplification." [13]

Descartes' reply invokes the so-called traditional theory of reference that Kripke and others have sought to overturn. Ordinarily, that theory is expressed with respect to language: a piece of language (a designator, in Saul Kripke's terminology) has its reference or denotation determined by its sense or connotation. But the theory can be extended as well to the reference or intentionality of thoughts, on which the referentiality of language is often thought to depend. The point is that there is something qualitative, expressed by the designator and existing in the mind, that determines reference. For Descartes, that qualitative something is an idea. His version of the theory is a very strong one, for the idea represents, in the sense of presenting to the mind, the essence of what is referred to. The thesis of the indivisibility of essences amounts to the claim that the idea gives us either the whole of the essence of a thing or none of it, and that unless the idea gives us the whole of the essence of a thing, the idea is in no sense an idea of that thing. In the case under discussion, only if an idea represents the essence of God can we by means of that idea think or talk about, or in any way refer to, God. To change the idea that in fact represents the essence of God, either by adding to or taking away from it, as Descartes puts it, is ipso facto to make it impossible to think about or refer to God by means of it, for it would then give us a different essence and thus be the idea of something other than God—of Pandora, for example. Hereafter, we can refer to this version of the traditional theory by saying simply that essence determines reference.

The components of Descartes' version of the traditional theory are to be found not only in Descartes' response to Gassendi. His position is at least adumbrated at the outset of the *Discourse on Method,* where he argues that reason is equal in everyone. It is reason alone, he says, that sets us apart from the animals and makes us human; he is "inclined to believe that it exists whole and entire in each of us, and that in this view [he] follow[s] the common opinion of the philosophers, who say that there are differences of degree only between accidents, and not between the forms (or natures) of individuals of the same species." [14] In dealing

13. AT 7:371; CSM 2:256.
14. AT 7:2–3: CSM 1:111–12 (slightly modified).

with forms or natures, that is, with essences, it is all or nothing. This exclusive dichotomy applies to knowledge of essences as well. In part 2 of the *Discourse,* Descartes himself is prepared to acknowledge the boldness of his claims to advances in geometry and algebra. His apology is that even a child who properly does a certain sum has learned all there is to be learned about that sum. "There being but one truth for each thing, anyone who finds it knows as much as one can know about that thing." [15]

The referentiality of language clearly depends on the referentiality of thought. In perhaps his best-known statement on the nature of ideas, Descartes gave a definition of the term *idea* at the outset of the geometrical proofs appended to the Second Replies: "I understand this term to mean the form of any given thought, immediate perception of which makes me aware of the thought. Hence, whenever I express something in words, and understand what I am saying, this very fact makes it certain that there is within me an idea of what is signified by the words in question." Unless I have the idea of a given thing, I cannot talk about that thing—at least not in a way that I would understand. And presumably, if others understand what I say, then they will have the idea. [16]

The same view is espoused in a letter of July 1641, the period during which Descartes was working on his *Replies.* Mersenne had conveyed to him a letter, the authorship of which is unknown, that questioned what Descartes meant by the ideas of God, the soul, and insensible things. [17] In his rather abusive reply, Descartes asked whether it was "credible that [the author] was unable to understand what I meant by [these ideas], since by them I mean nothing else but what he himself necessarily had to understand when he wrote you that he did not understand." Mersenne's correspondent could not have understood the terms *God, soul,* and *insensible things,* according to Descartes, without understanding what Descartes and everyone else mean by the ideas of them. Descartes then issues a reminder that ideas are not images depicted in the corporeal fantasy; rather, he "generally calls by the name 'idea' everything that is in our mind when we conceive a thing, in whatever way we conceive it." [18]

This conception of an idea is picked up in Descartes' definition of the

15. AT 4:21; CSM 1:121 (modified).
16. AT 7:160; CSM 2:113. Thus in the *Discourse* Descartes distinguishes people from animals on the basis of their speech. Because they lack thought, animals lack ideas, and thus even if they can parrot words "they cannot speak as we do: that is, they cannot show that they are thinking about what they are saying" (AT 6:57; CSM 1:140).
17. AT 3:375–77.
18. AT 3:391ff. I am grateful to David Behan for drawing this letter to my attention and for discussing it and other issues of this essay with me.

notion of objective reality in the geometrical proofs of the Second Replies, which connects intentionality with the traditional theory of reference. "By [objective reality of an idea] I mean the being of the thing which is represented by an idea, insofar as this exists in the idea. . . . For whatever we perceive as being in the objects of our ideas exists objectively in the ideas themselves."[19] We think about a thing just in case we have it in mind, that is, just in case we have its idea or essence in mind.

Finally, in his letter to Clerselier responding at least to the condensed version of Gassendi's rebuttals, Descartes invoked his version of the traditional theory against the objection that "not everyone is aware of the idea of God within himself." If such were true of someone, that person could not talk about God or understand those who did. Probably with full intention, Descartes evokes Anselm's atheistic fool: "The only way of denying [that we have an idea of God] would be to say that we do not understand the meaning of the phrase 'the most perfect thing which we can conceive of'; for this is what everyone calls *God*." Someone without such an idea "is saying not only that he does not know God by natural reason, but also that neither faith nor any other means could give him any knowledge of God. For if one has no idea, i.e., no perception which corresponds to the meaning of the word 'God,' it is no use saying that one believes that *God* exists. One might as well say that one believes that *nothing* exists." According to Descartes, to say that God exists without the idea of God, in his sense of the term, is not to know what one is talking about. "[I]f we take the word 'idea' in the way in which I quite explicitly stated I was taking it, . . . then we will be unable to deny that we have some idea of God."[20] As the text makes clear, the condition is not only sufficient but also necessary.

Gassendi's Rebuttals

Typically, Gassendi's text in the Rebuttals on just the issue of the indivisibility of essences and its connection with the idea of God runs to three pages *in folio* and therefore to about fifteen times the length of Descartes' text that he was rebutting.[21] Also typically, this material tends to be rather

19. AT 7:161; CSM 2:113–14. Nadler's formulation of this notion clearly suggests the traditional theory: "When we have the idea of an external object, certain properties of the object (those picked out by the idea) which it possesses 'formally' are thereby 'objectively' in the mind" (1989, 158–59). Descartes' view seems to be somewhat stronger, however, requiring not just certain properties of the object but its essence.

20. AT 9-A:209–10; CSM 2:273.

21. Gassendi [1658] 1964, 3:352b–355b. The French translation of the *Disquisitio* by Bernard Rochot (Gassendi 1962) is very useful in dealing with this work.

unwieldy and diffuse, but the thrust of Gassendi's rebuttal may usefully be summarized under four headings: difficulties in the Cartesian idea of God; an ad hominem argument against Descartes; an argument against essences; and difficulties for all Cartesian ideas, which Gassendi finds under the metaphysical microscope.

Gassendi raises several difficulties for the idea of God that Descartes claims to have. The difficulties suggest that he himself is arguing from either agnosticism (or at least blind fideism) or else outright atheism.[22] Sometimes the difficulties are philosophical, as when Gassendi argues that the essence of God cannot be both infinite and indivisible; for if God is extended, he is not indivisible, but if he is not extended, he is not infinite. And he argues (proleptically invoking a doctrine of internal relations in G. E. Moore's sense) that indivisibility of essence is not a matter of simplicity but of an indissoluble connection among parts—for example, between rationality and animality in man—which in the case of God is not known. Sometimes the difficulties are technical theological ones resulting from taking anything to be the essence of God *tout court*. Gassendi motivates his discussion of this sort of difficulty with a reference to Moses at Exodus 33:22–23: "And it shall come to pass, while my glory passeth by, that I will put thee in a cleft of the rock and will cover thee with my hand while I pass by. And I will take away mine hand, and thou shalt see my back parts: but my face shall not be seen" (KJV). That is, Moses, who spoke with God, nonetheless saw him only "through a glass, darkly" and did not know his essence.[23] The secular flavor of Gassendi's strategy is better expressed, however, by an allusion he makes to a text of Cicero's:

> Inquire of me [Cotta] as to the being and nature of God, and I shall follow the example of Simonides, who having the same question put to him by the great Hiero, requested a day's grace for consideration; next day, when Hiero repeated the question, he asked for two days, and so went on several times multiplying the number of days by two; and when Hiero in surprise asked why he did so, he replied, "Because the longer I deliberate the more obscure the matter seems to me."[24]

Following Descartes, says Gassendi, there would be no need for such delay in arriving at clear knowledge of the essence of God.[25]

---

22. Thus, not incidentally, arise the two principal interpretations of *libertinage érudit* offered by Popkin on the one hand and Pintard on the other. See Popkin 1979, chap. 5; and Pintard 1943, vol. 1, pt. 2, chap. 1, and pt. 3.

23. Gassendi [1658] 1964, 3:354b. See 1 Corinthians 13:12.

24. Cicero, *De natura deorum* 1.22; Cicero 1933, 59.

25. Gassendi [1658] 1964, 3:354a, 374.

Right at the outset of his attack on Descartes' doctrine of indivisibility, Gassendi draws attention, in a kind of ad hominem argument, to a reply Descartes had made to one of his objections against the Second Meditation. Recall the point of the piece of wax example in the Second Meditation: we know the piece of wax, not through the senses, not through the imagination, but by the mind alone *(mentis inspectio)*. Gassendi reads this to mean that we are supposed to know the wax itself, the substance of the wax or its essence, and he objects that in fact all we know of it are its accidents, that we cannot conceive of the wax apart from any extension, figure, and color.

> When you . . . say that the perception of color and hardness and so on is "not vision or touch but is purely mental scrutiny," I accept this, provided the mind is not taken to be really distinct from the imaginative faculty. You add that this scrutiny "can be imperfect and confused or perfect and distinct depending on how carefully we concentrate on what the wax consists in." But this does not show that the scrutiny made by the mind, when it examines this mysterious something that exists over and above all the forms, constitutes clear and distinct knowledge of the wax; it shows, rather, that such knowledge is constituted by the scrutiny made by the senses of all the possible accidents and changes which the wax is capable of taking on. From these we shall certainly be able to arrive at a conception and explanation of what we mean by the term "wax"; but the alleged naked, or rather hidden *[occultatam]*, substance is something that we can neither ourselves conceive nor explain to others.[26]

Descartes' reply is, perhaps, surprising: "I have never thought that anything more is required to reveal a substance than its various attributes; thus the more attributes of a given substance we know, the more perfectly we understand its nature. Now we can distinguish many different attributes in the wax; one, that it is white; two, that it is hard; three, that it can be melted; and so on."[27] However surprising, it is a view that he later seemed to repeat in the *Principles of Philosophy,* when discussing how "substance itself" is known.[28] The view seems to entail that our ideas can

26. AT 7:273; CSM 2:190–91.

27. AT 7:360; CSM 2:249.

28. Descartes, *Principles of Philosophy* 1.52: "We cannot initially become aware of a substance merely through its being an existing thing, since this alone does not of itself have any effect on us. We can, however, easily come to know a substance by one of its attributes, in virtue of the common notion that nothingness possesses no attributes, that is to say, no properties or qualities. Thus, if we perceive the presence of some attribute, we can infer that there must also be present an existing thing or substance to which it may be attributed"

be enlarged upon and that the essences of things are not indivisible, and, moreover, that our knowledge is limited to the accumulation of accidents or appearances. Thus, in order to rebut Descartes' reply to his objection concerning the idea of God, Gassendi merely quotes his reply to the objection concerning the idea of the wax.[29]

As is suggested by what Descartes has to say about the wax, the attributes that Descartes knows are, according to Gassendi, only the accidents of things *(adjuncta solum, sive accidentia)*.[30] We do have ideas of things, but these ideas are not of the sort that Descartes thinks they are. Indeed, we have ideas insofar as we know things *(ideae solum sunt rerum, quatenus illas novimus)*, but ideas are in our mind as pictures or images of things, and as in a portrait, the better the detail, the better the picture.[31] Image-ideas give us the accidents, not the essence, which we, as it were, obscurely suppose or conceive *(quasi suspicamur vel concipimus)* as lying beneath them. It is important to note that, like Locke on substance, whose language is notably similar, Gassendi does not commit himself to the existence of anything lying under the accidents. The same indeterminacy is found in what Gassendi has to say about the Scholastics' dictum that the essences of things are indivisible. "I shall not here inquire into the extent to which [their dictum] should be admitted, nor whether they know the essences to which they apply it." [32] Gassendi's claim can be read in at least two ways: there are essences, but knowledge of them is questionable, or knowledge of them is questionable because whether there is anything to be known is questionable.

Gassendi clearly understands Descartes to equate the idea of a thing with its essence *(aequi parare . . . ideam essentiae)*.[33] As he also puts it, for Descartes there is no difference between the being of a thing in itself and its being known by another *(nihil differre esse rem in se et cognosci ab aliquo)*, with the result that the essence does not contain anything that the "true" idea does not also contain.[34] According to Gassendi, this linkage just begs the question whether there are such ideas. Our initial idea

---

(AT 8-A:25; CSM 1:210). I say *"seemed* to repeat," for a distinction must be made here. In the later text, Descartes may rely on qualities for knowledge of *existence* rather than of *essence.*

29. Gassendi [1658] 1964, 3:352b–353a.
30. Ibid., 353a.
31. Ibid., 353b.
32. Ibid., 353a.
33. Ibid., 355a.
34. Ibid., 354b.

of a mite, for example, contains much less than the idea of it that we get from a microscope, which reveals what appeared to be a mere point to have a tail, legs, and other parts. As for the mite, so mutatis mutandis for the Almighty, who under the metaphysical microscope, as we might say, is revealed to have perfections not contained in our initial idea of him.

Gassendi's examples also beg the question, but two points of interest emerge from his discussion of them. First, he suggests a connection between the issue of scientific realism—whether the entities posited by science really exist, say, as the essences of things—and the relativity of sense perception that is often used to argue skepticism with respect to such entities. Gassendi's point is that by viewing an object under a microscope or through a telescope, or just by closing the distance to it, as when we recognize an approaching friend, we alter our idea of the object. This point is of help in understanding how Gassendi views essences. Second, Gassendi objects that what is not initially noticed in an idea is not in it "because the idea is the measure, not of the thing, but of the knowledge had of the thing." [35] Now, in his letter to Clerselier, Descartes directly replied to none of Gassendi's criticisms of the indivisibility doctrine. As it happens, however, Gassendi had made the same objection in another context, and to it Descartes did reply. Here is the objection as Descartes relates it: "[E]ven if I find no extension in my thought, it does not follow that my thought is not extended, because my thought is not the standard which determines the truth of things." [36] Alas, Descartes in reply takes the claim that thought is the standard determining the truth of things to be only an assertion of the autonomy of reason: everyone should judge of things according to his own lights. "The most absurd and grotesque mistake that a philosopher can make is to want to make judgments which do not correspond to his perception of things." [37] Descartes' followers made more of the claim, however.

Ideas as the Measures of Things

Not surprisingly, Arnauld was concerned with what is required for belief specifically in God, and like Descartes he held the traditional view that reference is determined by essence. At least, the knowledge of certain es-

35. Ibid., 355a.
36. AT 9-A:207; CSM 2:272.
37. AT 9-A:208–9; CSM 2:273. The only point of interest in Descartes' reply is his incipient deism. "Even with respect of truths of faith, we should perceive some reason which convinces us that they have been revealed by God" (AT 9-A:208–9; CSM 2:273). Cf. *Essay* 4.18.6, in Locke 1975, 693.

sential attributes is necessary to achieve the reference to God required for proper belief in him. "The knowledge that the pagans had of the divinity was so imperfect, so confused, and so full of things unworthy of God, not only in the people but also in the most exalted philosophers, who had thoughts of God that destroyed his true and infinite being, that it might be said without difficulty that they were as ignorant of God as are atheists and that their knowledge was a kind of atheism."[38]

Whether Arnauld also asserts the indivisibility of essences and therefore of ideas is an open question, but it is a doctrine to which he is clearly committed. If Descartes begged off the claim that the idea is the measure of the thing, Arnauld erects it into the principle of all knowledge. In the Port-Royal *Logic* he says that the only alternative to destroying all certainty in human knowledge and establishing a "ridiculous Pyrrhonism" is to accept the principle that "everything contained in the clear and distinct idea of a thing can be truly affirmed of that thing."[39] There is no question here of asserting the autonomy of reason as Descartes had done in response to Gassendi. Rather, Arnauld makes the strong claim (however question-begging) that if this principle were not true, "we would have no knowledge of things, but only of our thoughts."[40] Thus arises the significance, not incidentally, of Arnauld's conception of an idea as "the thing itself" insofar as it exists in the mind, which makes the principle analytically true.[41] What is contained in the true idea of a thing is contained in that thing because idea and things are identical.

Malebranche, too, accepted the strong claim, but with a very different basis, which was in fact the gravamen of his long dispute with Arnauld. For Malebranche, ideas are not identical with things in the way thought by Arnauld, which would indeed make reason autonomous, but in a way contrary to all of Malebranche's Augustinian intuitions. Insofar as an idea is the thing itself in the mind, the mind's modifications are essentially representative, with the result that, contrary to Augustine, we are a light unto ourselves.[42] Instead, the idea of a thing is the essence in the mind of God according to which it is created. Malebranche thus makes explicit

---

38. *De la necessité de la foi en Jésus Christ*, pt. 1, chap. 7, in Arnauld 1775–83, 10:1908.

39. *L'art de penser*, pt. 4, chap. 6, in Arnauld 1775–83, 41:378.

40. Ibid., 379.

41. *Des vraies et des fausses idées*, in Arnauld 1775–83, 38:200.

42. "It would seem that Arnauld should be reminded of the words of Saint Augustine. 'Dic quia tibi lumen non es.' Do not maintain ... *that the modalities of your soul are essentially representative*" (*Réponse* 7, in Malebranche 1958–69, 6–7:63). In just these terms, Jolley has investigated the difference in the theories of ideas between Descartes and Leibniz on the one hand and Malebranche on the other. See Jolley 1990, esp. chap. 1.

the connection between indivisibility of essences and immutability of ideas. Because of the necessary connection between exemplar and example, everything in the idea of a thing can be asserted of that thing, and because of our access to these ideas, skepticism is overcome.[43] These ideas are in fact the divine essence itself, with the result that the ideas of things must be immutable, which Malebranche, citing Augustine, asserts in so many words in the Tenth Elucidation appended to the *Search after Truth*.[44]

In his *Examination of . . . Malebranche,* Locke quoted Malebranche's claim here about the immutability of ideas, only to find it incomprehensible. "For how can I know that the picture of any thing is like that thing, when I never see that which it represents? For if these words do not mean that ideas are true representations of things, I know not to what purpose they are." If this is not what Malebranche means, then, according to Locke, he must mean only (as Bishop Butler was famously to point out) that everything is what it is and not something else: "Thus the idea of a horse, and the idea of a centaur, will, as often as they recur in my mind, be unchangeably the same; which is no more than this, the same idea will be always the same idea; but whether the one or the other be the true representation of anything that exists, that, upon his principles, neither our author nor anybody else can know."[45]

Now, it may be that the problem here is Locke's inability to understand and that his criticism is basically irrelevant.[46] But it may also be that Locke's failure to engage Malebranche is evidence that the difference between them on this issue is so deep that direct argument is all but impossible. That is, difference on this issue points to systematic differences that are accepted or rejected as of a piece. It seems to me, in fact, that Locke takes up the same position against Malebranche that Gassendi had argued against Descartes.[47] Indeed, this whole dialectic extends even to our own time. A brief indication of one such extension will lead back to the issue of indivisible essences as debated in the *Objections* and *Replies*.

43. *Réponse*, in Malebranche 1958–69, 8–9:924–26.

44. Malebranche 1958–69, 3:130; Malebranche 1980, 613. See Augustine, *De libero arbitrio*, bk. 2, chap. 8ff.

45. Par. 51, in Locke 1823, 9:250. Locke later makes the same point against the same view, as found in the work of Malebranche's disciple, John Norris: "What wonder is it that the same idea should always be the same idea? For if the word *triangle* be supposed to have the same signification always, that is all this amounts to" (*Remarks upon Norris,* par. 23, in Locke 1823, 10:257). See Norris 1690, 208–9 (par. 33).

46. Such is the view of Jolley (1990, 192–93).

47. This is one way of expressing the central theme of my book *Battle of the Gods and Giants*.

Anticipations of Kripke

Two decades ago J. L. Mackie published a short paper in which he argued, as he later put it, that "Locke made but put aside a discovery about an ordinary use of language."[48] The "discovery" was no less than the discovery later made independently by Kripke, namely, that the traditional view of the reference-essence connection can be short-circuited in favor of direct reference in ignorance of the essence of what is being referred to. The short of the story goes as follows. Locke thought that we are ignorant of the real essence of gold, that is, the internal constitution on which its perceived properties depend. When we give its essence as something that is malleable, yellow, and so on, we are giving only a nominal essence of our own construction—basically a constrained convention about our use of the word *gold*. But since for obvious reasons it would be better to have the real essence, we are irresistibly tempted to suppose that we are referring to the real essence when talking about gold. Our ignorance is thereby only aggravated, for we then literally do not know what we are talking, or attempting to talk, about. Thus we see both the discovery and the reason why Locke sets it aside.[49]

Locke's recommendation is that we restrict the term *gold* to its nominal essence, which, according to Mackie, would recommend a phenomenalist metaphysics. I believe such a metaphysics to be actually part of the Lockean story. But the linguistic practice of restricting reference to nominal essence would be, according to Mackie, only at the price of great inconvenience.[50] The alternative recommended by Mackie is the Kripkean causal theory, which requires a realist metaphysics, not just with respect to the status of the object (antiphenomenalism), but also of its constitution (antinominalism). The atomic number 79 on this alternative is something irreducibly shared by all instances of gold. Locke, for whom "all things that exist [are] particular," cannot have anything to do with such universals.[51] So what is going on here?

48. Mackie 1976, 93. See Mackie 1974.
49. Mackie cites Locke's *Essay* 3.10.19 (Locke 1975, 501) as evidence that Locke thought we can refer to the real essence of gold, at least when a sample of it is present: "[B]y this tacit reference to the real Essence . . . the word Gold can signify nothing at all, when the Body itself is away" (1974, 179). While bare reference may thus be possible, no assertion is ever possible. Anything we may say of it "will always fail us in its particular Application" (3.7.50; Locke 1975, 470).
50. Mackie's attempt to state the inconvenience just begs what is at issue, however. "What a nuisance it would be if a discoloured piece of gold could not be called gold, or if a sufficiently convincing counterfeit diamond had to be called a diamond" (1974, 179).
51. *Essay* 3.3.1, in Locke 1975, 409.

First, Locke is not the first anticipation of Kripke. The skeptic Simon Foucher argued essentially the same position in 1675 against Malebranche. The context is the following dilemma: either resemblance is required for representation or it is not; if resemblance is required for representation, then either we know nothing of material things or the essences of soul and matter are not different; if resemblance is not required for representation, then the senses are capable of representing material things.[52] Foucher's own position is to grasp the first horn, with the skeptical disjunct of the consequent. Resemblance is required, according to Foucher, despite the Cartesian analogy of words that are able to represent without resembling what they represent. *Tree* does not resemble a tree, yet we are able to think of a tree by means of *tree,* and as for *tree* so for the idea of a tree.[53] Foucher accepts the traditional theory of reference at least to the extent that, "properly speaking, it is not the word that represents the tree, but the idea this word excites in us which does so."[54] Now comes the anticipation. "I do not say that the image [read: *idea*] that this word excites in us is like what the tree is in itself; but I do say that it is like the effect that this object produces in us through our senses. Nor do we maintain that this word represents any other thing to us. *For it would still have its full meaning were we in complete ignorance of what the tree is in itself (as is certainly the case),* since this word allows us to know nothing more than when we actually see the object."[55] Although we must have some idea of the tree in order to talk about it, that idea does not give us its essence.

Nor was Foucher the first to argue this position. Once expressed in the terms above, Foucher's position can be seen in Gassendi's original objection to Descartes. Recall Gassendi's claim about the wax: "From these [accidents and changes of which the wax is capable] we shall certainly be able to arrive at a conception and explanation of what we mean by the term 'wax'; but the alleged naked, or rather hidden, substance is something that we can neither ourselves conceive nor explain to others."[56] Indeed, the whole skeptical tradition emerges as an anticipation of Kripke.

52. Foucher [1675] 1969, esp. 51–52. See Watson 1966, esp. chap. 4.

53. Foucher does not cite any Cartesian text, but see *Le monde:* "[W]ords, having no resemblance to the things they signify, still make us conceive of them" (AT 11:4; CSM 1:81 [slightly modified]). The distinction between nonresembling sensations, which signify, and resembling ideas, which represent, is one that Foucher will not admit.

54. Foucher [1675] 1969, 57.

55. Ibid., 57–58; emphasis added. Note that Foucher is not anticipating later cluster theories, which agree with the traditional theory that we can, indeed must, know what we refer to.

56. AT 7:273; CSM 2:190–91.

For the traditional skeptic does not allow knowledge of the essences of things and yet does not deny that reference to them is possible.[57] There is an important difference, however, between Gassendi and the skeptics that must now be investigated in some detail. It is a difference that relates to the proper interpretation of Locke and, more important, to Descartes' doctrine of the indivisibility of essences.

Gassendi's Neutral Nominalism

In his *Instititio logica*, Gassendi takes over Sextus' semiotic of indicative and telling signs.[58] A telling sign *(signum commonefactivum)* signifies its natural accompaniment; for example, smoke is a telling sign of fire, and milk of pregnancy. Both Gassendi and the skeptic accept telling signs as a means of not getting burned. An indicative sign *(signum indicativum)* is such that it could not exist unless the thing signified also existed. The indicative sign was rejected by the skeptics for it was used by dogmatists to argue from the appearances of things to their reality or essence. Gassendi, however, accepts the indicative sign, and the example he gives of it shows why. Just as at the macro level there must be passages in a body if another body is able to pass through it, so at the micro level of the skin, for example, there must be pores to allow it to transpire sweat. The point is that no new kind of entity such as an essence is thereby introduced. Gassendi thus restricts himself to "transduction."[59]

The restriction is confirmed by his tripartite division among things that are hidden from us *(res . . . occultae)*. First of all, manifest things are known by themselves—for example, "the light of day or that it is day, or the external aspect of things *[externa rerum facies]* that spontaneously strikes our eyes and discloses itself to sight without any obstructing veil between."[60] Things *temporarily* hidden are those which, though naturally manifest, are circumstantially hidden: e.g., a fire hidden by a building, but which may be knowable by a telling sign. The *naturally* hidden are knowable only through inference: the pores of the skin are inferred from indicative signs. Finally, the *totally* hidden are utterly beyond our intellect: we cannot know whether the number of the stars is even or odd. Even in this last category, there is no question of some essence lying beyond our

57. A relevant exception is Gorgias. But he held that *all* reference is impossible because nothing exists and a fortiori nothing exists to be referred to.

58. *Instititio logica* 1.81. For the fuller version of the story, see Lennon 1991.

59. See McMullin 1978, 15–16, for the terminology recently used to distinguish this kind of induction.

60. *Syntagma philosophicum*, in Gassendi [1658] 1964, 1:68b.

powers of comprehension beneath the appearances of things. Indeed, it makes perfect sense to talk of the number of the stars as even or odd, and one of those who disagree on the question is even right about it. I am suggesting that with respect to the sort of essences admitted by Plato, or Aristotle, or Descartes, Gassendi is no skeptic who suspends judgment as to what they might be. He believes that there are no such things at all.

The result sheds some light on what is found under the microscope, both physically and metaphysically. Recall Gassendi's point that by viewing an object under a microscope or through a telescope, or just by moving closer to it, we alter our idea of that object. The clear suggestion is that by improving the circumstances under which a thing is perceived, our knowledge of it is made more adequate. An important question is whether in principle that knowledge can ever be perfectly adequate and what that knowledge would be of. In Lockean terms, what would we see with microscopical eyes as ideas of secondary qualities of a thing finally fell away? For Locke, we would see primary qualities of the ultimate particles composing that thing, namely, its real essence. This seems to be, if not Gassendi's answer, then at least one that is anticipated by what he says.[61]

In these terms, what one would expect from Gassendi and Locke is that real essences are fixed; for these just are the (constitutions of) things. At any given time they are certainly fixed, being just what they are and nothing else. Nominal essences that coincide with real essences should also be fixed, for these are abstract ideas, which are entirely of our own fabrication. These are the basis for Locke's version of the doctrine of the immutability of essences, which "is founded on the Relation between them, and certain Sounds as Signs of them; and will always be true, as long as the same Name can have the same signification."[62] Finally, nominal essences different from real essences would, through Hillary Putnam's "division of

61. As Grene puts it, for Gassendi, "essences are not known: Locke's nominal essences in preview. In particular, . . . Gassendi insists that we cannot know substances. Or as Locke will put it, our knowledge of them reaches only a very little way" (1985, 162). My slight departure from her position would be to say that substances are not at all known, but essences (Locke's "nominal essences in preview") are known only a little way. In any event, Gassendi rejects the view that beneath subjective appearances apprehended through sensation (data), there is something different in kind from them that is known through intellection (theory). In what may or may not be a kind of phenomenalism, things just appear different under different conditions. Gassendi thus undercuts the skeptic's argument that if a thing is *a*, then contrary to fact it would always appear *a* to everyone. A thing that in itself is *a* in this sense is inconceivable to Gassendi.

62. *Essay* 3.3.19, in Locke 1975, 420.

linguistic labor," gradually be aligned with them, with fixedness as an ideal limit. Accuracy of reference would of course vary with the alignment. The idea is that for most purposes we would refer to (what we call) gold by means of nominal essences (color, etc.). In cases of dispute—for example, whether a given object was "really" gold—we would call upon chemists, the experts whose testimony would gradually alter our nominal definitions, gradually aligning them with the real constitutions of things.

Although this picture is not Locke's, at least not nominally—the real picture, if you will, might look rather different—it does focus much of what Gassendi says.[63] It would seem that according to him there is no difference between the physical and metaphysical microscopes. For according to him, what we find under the physical microscope is just more of the same. Contrary to Mackie's phenomenalist interpretation of Locke, for example, this is, initially at least, an ontologically neutral picture. The items that change from the macroscopic to the microscopic levels might be constituents of a phenomenal object, but they might also be appearances of a nonphenomenal object. They are not, however, appearances of an essence different from them in ontological kind.[64] For in these terms there is an ultimate "appearance" that just is the object. What we see are the ultimate constituents, the "atoms," that compose the world and explain—in the same terms used at the macro level—everything else, as pores in the skin explain the transpiration of sweat. The relativity of sense perception that is cited by skeptics against knowledge of essences just ceases at this point—not in the intellectual apprehension of essences (there is no such apprehension, and there are no such things to be apprehended), but in the sensible apprehension of the ultimately real, which is no longer theoretical but observational. This realist reading differs from

63. Alas, only in the case of abstract ideas coinciding with real essences does Locke himself confirm our expectations. We get just the opposite of what we would expect in the two other cases: nominal essences differing from real essences and real essences themselves. "*Essences* being taken for *Ideas*, established in the Mind, with Names annexed to them, they are supposed [!] to remain steadily the same, whatever mutations the particular Substances are liable to." Thus Alexander and Bucephalus may perish, but supposedly not the essences of their species, Man and Horse. In the case of real essence, we expect them to be fixed, but Locke tells us that they change. "All things, that exist, besides their Author, are all liable to Change; . . . In all which . . . Changes, 'tis evident, their real *Essence, i.e.,* that Constitution, whereon the Properties of these several things depended, is destroy'd, and perishes with them" (*Essay* 3.3.19, in Locke 1975, 419).

64. Grene writes, "To hope radically to transcend appearances and to grasp intellectually the inner nature of things is as [Gassendi] puts it . . . 'no more preposterous than to wish to have wings or to remain forever young.' Philosophically, this is [Descartes' and Gassendi's] deepest incompatibility" (1985, 158).

the phenomenalist reading that construes appearances as constituents in that the realist reading makes what is epistemologically primary into what is also ontologically primary. The appearance of a thing deriving from its atomic structure explains all else because it is the ultimately real.

Descartes' Ontology

None of this, of course, is to be found in Descartes. The source of Descartes' language for expressing his doctrine of the indivisibility of essences and the theory of reference that depends on it comes as no surprise. A principal source for his views of "the philosophers" is Eustachius a Sancto Paulo, in the fourth part of whose *Summa philosophiae* is a text that Descartes may have closely paraphrased: *"nihil pertineat ad essentiam addi vel detrahi potest quin statim mutetur, & fiat alia natura seu essentia."*[65] But despite the Aristotelian-Scholastic source for the language expressing his view, his view itself differs radically from that source. For example, although Descartes told Gassendi that "the substance of the wax is revealed by means of its accidents," scholars today should not read his claim in the Scholastic terms that Gassendi did. First, while the accidents of the wax's substance reveal that substance, the piece of wax itself is not that substance. Rather, the piece of wax is an accident of the single material substance, *res extensa*. In more properly Cartesian language, it is a mode of extension, that is, a way in which extension exists (thus the French *façon d'être* as a translation of the Latin *modus*).[66] Second, the connection between substance and mode is not the connection of inherence in the Aristotelian substance ontology. Instead it is a deductive connection, such that extension *(extensio* or *res extensa)* stands to individually extended things *(extensa)* as the axioms and postulates stand to the theorems of Euclidean geometry.[67] If this deductivism makes sense, it helps provide an understanding of how a fuller version must go of Descartes' reply to Gassendi on our clarification or explication of the idea of God.

At the lowest level of comprehension is the idea of God we necessarily

65. Eustachius a Sancto Paulo 1620, 21. Gilson (1979, 103–4) cites pages 34–35 for the first edition, of 1609.

66. One particularly relevant argument for this interpretation is based on the infinite, or at least indefinite, divisibility of matter. See, e.g., *Principles of Philosophy* 2.34. If matter and extension are really identical and extension is the essence of matter, then if Descartes is not to contradict himself on the indivisibility of essences, he must mean that extended things *(extensa)* such as the piece of wax are divisible, not their essence *(extensio)*.

67. See Bracken 1974.

have as conscious beings.[68] This is the reflective innateness that Robert McRae distinguishes. In addition to the idea of God, "certain universal concepts such as 'thought', 'existence', 'thing,' . . . are all capable of being derived by intuitive induction from my experience or consciousness of *any* individual act of thinking. Every man has an implicit knowledge of these concepts from the mere fact that he thinks and is conscious of thinking." [69] Such an idea is the basis for the more perfect knowledge open to us, not only of God but also of geometry, for example. Recall that in his initial reply to Gassendi, Descartes explicitly assimilated the two cases: "[S]imilarly, the idea of a triangle is not augmented when we notice various properties in the triangle of which we were previously ignorant." In this sense, relieving ignorance both of geometry and of God consists in making explicit through deduction what was previously only implicit.[70]

It is interesting to speculate about even higher levels of comprehension. The beatific vision of God enjoyed by the angels and the blessed would be a single intuition of all that is contained in the idea of God, the intuition into which all deductions are in principle convertible. Once again the analogy with geometry stands. No one, not, at least, in this life, comprehends even a triangle in the sense of knowing all the theorems true of it. Finally, to distinguish angelic knowledge of God from God's knowledge of himself, we might distinguish passive from active or creative intuition. Once again the analogy with geometry is instructive. While angels immediately know all of geometry, their knowing it does not make it true. God's knowing it does. These speculations lead to two others.

Two Speculations

Descartes' doctrine of indivisibility applies to any essence and therefore is relevant to thought about or reference to any given thing, including

68. Caution is to be exercised with the term *comprehension*. On the basis of the *Conversations with Burman,* J.-M. Beyssade tries to sort out knowing the finite, indefinite, and infinite as objects of *comprehendere, concipere,* and *intelligere.* Ariew prefers Alquié's view that we know God without comprehending him but that *conceiving* is sometimes assimilated to *comprehending.* See Ariew 1987, 156ff. Whatever the terminology, there must be degrees of knowledge of God.

69. McRae 1972, 41–42. It is in this sense that the idea of God is imprinted in us in the way in which (as Descartes explains to Gassendi) "a picture whose inimitable technique showed that it was painted by Apelles could be said to carry the mark which Apelles imprinted in all of his pictures" (McRae 1972, 41–42).

70. Thus innate knowledge, in McRae's second sense of the term, is produced (1972, 42ff.).

God, whose idea Gassendi puts in question. It is natural therefore that Descartes should raise the issue of false gods. There is, however, an important issue that can be teased from the connection of the doctrine specifically with the idea of God. Only in one other text does the doctrine arise—as it happens, in the Sixth Objections and Replies. There it is put to Descartes that clear perception cannot determine our will and thus remove freedom of indifference, for "if indifference cannot be a proper part of human freedom, neither will it find a place in divine freedom, since the essences of things are, like numbers, indivisible and immutable." [71] The objection is that because the essence of the will is like all essences indivisible and everywhere the same, if the human will is not indifferent, then, contrary to the dogma that God's creation of the world is indifferent, neither is the divine will indifferent.

Descartes' reply is of great significance. The doctrine of indivisibility is irrelevant, he says, because "no essence can apply univocally both to God and His creatures." In this case, will in God and will in us are two different things. Indifference in us is a function of our ignorance—of our failure to be constrained by perception of the true or the good. But "the supreme indifference to be found in God is the supreme indication of his omnipotence." [72] Descartes seizes this occasion for only the second and last published version of a view that a decade earlier he had described to Mersenne as lying at the foundation of physics and metaphysics. [73] This is the doctrine of the created eternal truths. "It is self-contradictory to suppose that the will of God was not indifferent from eternity with respect to everything which has happened or will ever happen; for it is impossible to imagine that anything is thought of in the divine intellect as good or true, or worthy of belief or action or omission, prior to the decision of the divine will to make it so." [74] God did not will to create the world in

---

71. AT 7:417; CSM 2:281. The author(s) of this objection here tend to corroborate Eustachius as Descartes' source for the doctrine of indivisibility. In the text in which he sets out the doctrine, Eustachius refers to *Metaphysics* 8.3, where Aristotle says, "[A]s, when one of the parts of which a number consists has been taken from or added to the number, it is no longer the same number, but a different one, even if it is the very smallest part that has been taken away or added, so the definition and the essence will no longer remain when anything has been taken away or added" (1043b37–1044a2, Aristotle 1941, 816). This shows only that the author(s) connect the doctrine with Aristotle, which they conceivably could do independent of Eustachius, either directly or through some other author. More likely, Eustachius was their source and thus more likely Descartes' source as well.

72. AT 7:433; CSM 2:292. See Caton 1975.

73. To Mersenne, April 15, 1630, in AT 1:143.

74. AT 7:431–32; CSM 2:291.

time because it was good to do so; nor, says Descartes, did he will that the three angles of a triangle equal a straight angle because it was true. Rather, the converse is true in both cases.

In effect, we have just come full circle. The first published version of this doctrine occurred in Descartes' reply to Gassendi, who had thought it "very hard to propose that there is any 'immutable and eternal nature' apart from almighty God," as Descartes appeared to do at the outset of the Fifth Meditation.[75] In reply, Descartes announced his doctrine: "I do not think that the essences of things, and the mathematical truths which we can know concerning them, are independent of God. Nevertheless, I do think that they are immutable and eternal, since the will and decree of God willed and decreed that they should be so."[76] It was this text, according to Emile Brehier, that in fact led to the objection from the authors of the Sixth Objections.[77]

Why did Descartes subscribe to such an extreme version of voluntarism? The construal of eternal truths, or essences, as both uncreated and independent of God had at least since the time of Aquinas been conceived of as turning them into rival divinities in their own right. Thus Aquinas, to avoid this sort of polytheism, identified eternal truths, or essences, with the divine essence. The upshot was that in knowing anything, one to some extent knew God and thus everything. According to Brehier, the holism entailed by such a solution thus closed it off to Descartes, who required an atomist, proposition-by-proposition application of his criterion of clarity and distinctness in the acquisition of knowledge. Instead, therefore, Descartes construed essences as different from God but nonetheless dependent on him. Essences are indivisible, fixed, eternal, and immutable, but they are so because of the divine will.

Gassendi is no Platonist who runs the risk of this sort of polytheism because of his Platonism. Rather, the danger in this regard, at least from the Cartesian perspective, arises from his nominalism. The situation is not unlike that related by Tertullian, whose text is quoted and discussed by the church historian Eusebius. According to Tertullian, reports of the resurrection of Christ reached the emperor Tiberius, who seemed open to the doctrine and recommended it to the Roman senate, which rejected it for want of proper examination. This implies that whatever the basis for their decision, the emperor and senate were in a position to define deities into existence. Says Tertullian: "[Among them] godhead is conferred by human approval. If a god does not satisfy man, he does not become a

75. AT 7:319; CSM 2:221.   77. Brehier 1967, 193.
76. AT 7:380; CSM 2:261.

god." [78] It is to this Hobbesian, nominalist theism that Descartes objects when Gassendi indicates that we can alter our idea of God: "This is how the ideas of Pandora and of all false Gods are formed by those who do not have a correct conception of the true God." [79]

Why Pandora? Perhaps her name was drawn at random from classical mythology. But perhaps not. First, Pandora was a fabrication even in mythology. To revenge himself upon Prometheus ("Forethinker"), Zeus "ordered Hephaestus to make a clay woman, and the four winds to breathe life into her, and all the goddesses of Olympus to adorn her." [80] This was Pandora, who was ultimately taken as a wife by Prometheus' brother Epimetheus ("After-thinker"). Despite the warning from Prometheus to his brother against just this danger, Pandora soon opened a jar and thus unleashed the Spites (Old Age, Labor, Sickness, Insanity, Vice, and Passion, but also Hope), which stung her and Epimetheus and then afflicted the human race. Although Pandora was thus only too efficacious causally, she was still a being whose idea was *formed*.

But there may be an even more interesting version of this explanation for Descartes' choice. Because of her endowments, Pandora was generally thought in the ancient world be the "all-gifted." But Robert Graves points out that this is a mistake. "Hesiod's account of Prometheus, Epimetheus and Pandora is not a genuine myth, but an anti-feminist fable, probably of his own invention. . . . Pandora ('all-giving') was the Earth-goddess Rhea, worshipped under that title at Athens and elsewhere." [81] Thus does Descartes speak of the ideas of Pandora and of "all false Gods." That is, even in mythology there are true and false gods, and mistaking one for the other is the same kind of error to which Gassendi is liable with his theory of how the idea of the one true God is formed.

78. Eusebius 1965, 76–77.

79. Nor surprisingly, Plato has a text on reference, or at least on intentionality of belief, similar to Descartes' and with respect to essentially the same topic. There is no question here of belief in the one true God, obviously, but of the gods. Socrates in *The Apology* stands accused by Meletus of the propagation of belief in gods different from those recognized by the state and therefore of atheism. Socrates' response is that the charge is contradictory: as he puts it, he is to be "guilty of not believing in the gods, but of believing in the gods." He argues the senselessness of the charge on this basis, ultimately, that to believe in supernatural activities without believing in the supernatural and thus in the gods is impossible (26c–28a). There are difficulties in this text, but the only way to understand the argument is that, whatever *Theaetetus* may indicate on the problem, belief in *false* gods is impossible; either we believe in the true gods through their "activities" or we believe in none at all.

80. Graves 1955, 1:144–45.

81. Ibid., 148. See also Rose 1974, 23.

STEPHEN MENN

—————— ✃12✄ ——————

THE GREATEST STUMBLING BLOCK:
DESCARTES' DENIAL OF REAL QUALITIES

After his philosophical objections to the *Meditations,* Arnauld also gives a series of theological objections. He wants Descartes to say that the method of doubt does not apply to faith or morals, and he wants him to correct the apparent Pelagianism of the Fourth Meditation; but Arnauld predicts that "the greatest stumbling block for the theologians" (AT 7:217) will arise from Descartes' denial of sensible qualities. Descartes summarizes Arnauld's objections: "He thinks that my opinions do not agree with the sacrament of the Eucharist, because (he says) we believe by faith that when, in the eucharistic bread, the substance of the bread has been removed, the accidents remain there by themselves; but he thinks that I do not admit real accidents, but only modes, which cannot be understood, and indeed cannot exist, without some substance in which they are present" (AT 7:248). From a twentieth-century perspective, this objection seems surprising. In the first place, although Arnauld flatly asserts that "Descartes thinks there are no sensible qualities" (AT 7:217), it is unclear where Arnauld is getting this: Descartes does not seem to say anything like it in the *Meditations.* But supposing Descartes did say this, it seems strange that the only objection to it comes from a technical point of theology: surely we all believe, independent of theology, that we live in a world of colored and hot or cold objects, and surely our ordinary beliefs should put up some protest before succumbing to Descartes' arguments. But in fact nobody raises this kind of objection against Descartes: when Arnauld and Descartes and (later) the sixth Objectors (AT 7:417) debate the status of sensible qualities, the objections arise from Scholastic theology rather than from ordinary belief, and they are couched entirely in the language of substances and modes and real accidents.

Reading the texts of the debate, we may well feel that we no longer understand, as Arnauld and the sixth objectors did, what was at issue when Descartes denied the reality of sensible qualities. In this essay I want to elucidate what Descartes was denying, why Arnauld and the others were troubled by this denial, and why the Eucharist was at the center of their objections. In the process, I want to bring out why this denial was essential to Descartes' philosophical and scientific project. I will not discuss Descartes' *answers* to the objections, thus not, in particular, his eucharistic theology (here I defer to Armogathe's *Theologia cartesiana*). I will be presenting a reading of Descartes, not of the objectors; but it will be a reading of Descartes very much from the perspective of the traditional philosophers and theologians of his own time, and I hope it will explain some of their responses.

To understand what Descartes was saying and what Arnauld was reacting to, we must first recognize and reject the implicit assumption of most twentieth-century interpretations of Descartes: that when Descartes said that heat is not a real quality in fire, he meant that fire is not really hot. We can see that this assumption must be wrong from what Descartes says about figures. Descartes believes that bodies really have certain figures and that any scientific account of bodies will refer to these figures. He also believes that figures are qualities; but he denies that figures are *real* qualities, and he thinks that, in this denial, he is merely following a philosophical consensus. In a letter to Mersenne on projectile motion, Descartes starts by laying down some principles of his physics: in particular, he says that he "does not admit any real qualities" and that for this reason he does not "attribute any more reality to motion, or to all these other variations of substance that are called 'qualities,' than the philosophers [i.e., the Scholastics] commonly attribute to figure, which they call not a *qualitas realis* but only a *modus*." (AT 3:648–49; Descartes cites the Scholastic terms in Latin, in a French context). So for the Scholastics (as Descartes interprets them) some qualities are real qualities, and other qualities are only modes; figure is a quality (indeed, it is one of the four main species of quality that Aristotle recognizes in the *Categories*), but it fails to count as a real quality because it is only a mode. Descartes' interpretation of the Scholastics is in fact correct, at least for Suárez, who says that while some categories contain only *res* and some contain only modes, some contain both *res* and modes, "as in the genus of quality there are both heat and figure," heat being a *res* and figure a mode.[1] But the crucial point here is to note Descartes' attitude toward this Scholastic

---

1. Suárez, *Disputationes metaphysicae*, disp. 39, sec. 2, par. 17. (All references to Suárez in this essay are to the *Disputationes metaphysicae*.) Suárez notes that Aristotle listed figure last among the four species of quality (in *Categories* c.8) because it is "the lowest of them all, both in perfection and in way of being" (disp. 42, sec. 5, par. 9). That figure is merely a mode (but still truly a quality) is mentioned in Disputation 32 (sec. 1, pars. 14 and 19), Disputation 16 (sec. 1, par. 21), and elsewhere (in Disputation 42 [sec. 4, par. 15] *figure* is

position: he thinks the Scholastics are right about figures but wrong in regarding some other qualities as real, when in fact no other qualities have any more reality than figures do. In this letter, Descartes is concerned chiefly with motion—"motion [is] not a real quality, but only a mode" (AT 3:650)—but he wants his point to apply to *all* qualities, including sensible qualities: "[H]eat and sounds, or other such qualities, give me no difficulty, for these are only motions that are produced in the air" (AT 3: 649–50). So Descartes is not holding up figure and motion as examples of real qualities and arguing that sensible qualities have less reality; on the contrary, he is taking figure as an example of a quality that is *not* real and arguing that motion and sensible qualities are not (as the Scholastics believe) more real than this.[2]

This and similar texts show that for Descartes, as for Suárez, a quality can really belong to something, and be really a quality, without being a real quality: Descartes is using *real* consciously and precisely as a technical term. A real quality is a quality that is a *res;* something can fail to be a *res,* even though it is the subject of true predications, if it is a mode or an *ens rationis cum fundamento in re.*[3] When Descartes speaks of something's degree of "reality," he means the degree to which it is a *res:* "I

---

defined as "a mode resulting in a body from the termination of a magnitude"), but never defended at length. The division of qualities, with figure as the fourth species, was universal enough that it is recorded even in the Port-Royal *Logic* (Arnauld 1775–83, 41:137). Contrary to what Descartes seems to suggest, however, Suárez' thesis that figures are modes rather than *res* was not the most common Scholastic view (Suárez criticizes Pedro da Fonseca for taking a different position in Disputation 7 [sec. 1, par. 19]). In this essay I will present Scholastic positions from the point of view of the specifically Suárezian modification of Thomism; this almost always corresponds to Descartes' understanding of Scholasticism. In my essay "Suárez, Nominalism, and Modes" (Menn forthcoming) I give a much more thorough treatment of Suárez' position on the issues here discussed, putting him in the context of Spanish Golden Age Scholasticism and indicating his differences from other Scholastics on some of these issues.

2. On motion, compare *Le monde:* "[T]hey [the Scholastics] attribute to [even] the least of these motions a being much truer and more solid than they attribute to rest, which they say is merely the privation of motion. But as for me, I think that rest is just as much a quality that must be attributed to matter when it remains in one place, as motion is a quality that is attributed to it when it changes places" (AT 11:40). Although the Scholastics may have been unjust in considering rest as a mere privation (rather than a mode), Descartes' main point is to reduce motion to the level of rest, rather than to exalt rest to the level of motion. Thus in the *Principles of Philosophy* he argues that motion "is merely a mode [of the moved body], and is not some subsisting *res,* just as figure is a mode of the figured thing, and rest of the resting thing" (*Principles* 2.25).

3. In *Le monde,* in a context closely parallel to the text just discussed, Descartes writes that motion "follows the same laws of nature as generally do all dispositions and all qualities that are found in matter, those which the learned call *modi et entia rationis cum funda-*

have explained sufficiently how reality receives more and less, namely, that a substance is more a *res* than a mode is; if there are any real qualities or incomplete substances, they are more *res* than modes are, but less than complete substances; finally, if there is an infinite and independent substance, it is more a *res* than a finite and dependent one" (AT 7:185). A real quality (or a real accident in any category) would be a *res* really distinguished from its subject, that is, distinguished from it as one *res* from another *res*. Since (for Descartes, as for the Scholastics) any *res* other than God must be something created by God, and since God is free to create any one *res* without creating any other *res,* it follows that any real accident could exist separately from its subject (and from any other subject).[4] It is in this sense that Descartes denies the reality of qualities and other accidents: "I do not suppose any real qualities in nature, which would be added to substances (like little souls to their bodies), and could be separated from them by divine power" (AT 3:648).

Since *real quality* has this precise technical sense, Descartes' denial that colors, heat, or figures are real qualities does not commit him to denying that these are really qualities or that they really belong to things. In fact,

---

mento in re (modes and beings of reason with foundation in the thing) as well as *qualitates reales* (their real qualities), in which I naively admit that I find no more reality than in the others" (AT 11:40; here Descartes first cites the Scholastic terms in Latin, and then translates them into French). So Descartes recognizes that the Scholastics distinguish three ontological levels at which a quality might be located: real qualities are realer than modes, which are realer than *entia rationis cum fundamento in re,* and even these are not nothing *simpliciter,* but have some diminished kind of existence. Descartes' rhetoric is slightly overdone: he wants merely to reduce the alleged real qualities to the level of modes, not to reduce all these items to the level of *entia rationis.* An *ens rationis* is either a negation (or privation) or a *relatio rationis,* unless it is a mere figment with no *fundamentum in re.* Strictly speaking, although modes can properly belong to the category of quality (and to some other categories), no mere *ens rationis* can be truly a quality, or belong to any other category of being, although blindness can improperly or "reductively" be assigned to the category of quality.

4. It is a maxim of post-1277 Scholasticism (resisted only by some hard-line Thomists) that God's omnipotence entails that he can create any *res* without any other *res,* even if in the ordinary course of nature the first *res* is causally dependent on the second, as an accident is on its substantial subject. Accidents do in fact subsist by themselves in the eucharistic species. Suárez makes it necessary and sufficient for a real distinction (that is, of a distinction of one *res* from another *res;* a distinction might be *ex natura rei* without being in this sense *real*) that the two terms can be separated either naturally or supernaturally (disp. 7, sec. 2, pars. 9–12, 22–27). Descartes accepts the same criterion (*Principles* 1.60), with the same foundation in God's omnipotence. A real accident would be a *res* really distinct from its subject; when "we" followed the Scholastic view, we believed in "various qualities of bodies, as weight, heat, and others, which we imagined to be *real,* that is, to have an existence distinct from that of the body, and in consequence to be substances, even though we named them 'qualities'" (AT 3:667).

Descartes admits all of these as qualities of bodies: he says that the modes of a substance can equally be called its qualities (*Principles* I, 56),[5] and these include "heat and other sensible qualities" (*Principles* IV, 198). Descartes is equally liberal about forms, denying *substantial* forms (except the human mind) but accepting many other forms: *Principles* II, 23 says that "all variation in matter, or all diversity of its forms, depends on motion," *Principles* IV, 198 mentions the "form of fire," and *Le monde* not only speaks of the "forms" of the three elements but also criticizes the Scholastics for positing a prime matter "despoiled of all its forms and qualities" (AT 11:33).[6] There is no contradiction in any of this: just as a

5. "When we consider that a substance is affected or varied by them, we call them modes; when we consider that from this variation it can be denominated such-like [*talem . . . denominari*], we call them qualities" (*Principles* I, 56). This is a reference to the standard description of quality as *qua quales quidam dicuntur* (Suárez, disp. 42, sec. 1, par. 1, from the beginning of Aristotle's *Categories* c.8); this is neutral as to whether the qualities are *res* or not. A horse's whiteness *denominates* it "white" (rather than simply *nominating* or naming it), because the horse is named by the *denominative* term *white*, that is, not "whiteness" itself but a term grammatically related to it.

6. Descartes describes the forms of fire, air, and earth in *Le monde* (AT 11:26–29). Descartes thinks that "forms, at least the more perfect ones, are collections of many qualities, which have the power of mutually preserving each other" (AT 3:461); explaining the forms of the elements in *Le monde*, he says that "the forms of mixed bodies always contain in themselves some qualities that are contrary and harm each other, or at least do not tend to each other's preservation; whereas the forms of the elements must be simple, and not have any qualities that do not fit together so perfectly that each tends to the conservation of all the others" (AT 11:26). Answering the question of why he does not explain the elements using "the qualities called heat, cold, wetness, and dryness, as the philosophers do," Descartes says that "these qualities themselves seem to me to need explanation; and, if I am not deceived, not only these four qualities but all the others, and even all the forms of inanimate bodies, can be explained without needing to suppose for this purpose any other thing in their matter but the movement, size, figure, and arrangement of its parts" (AT 11:25–26). There is no question of *denying* these forms and qualities, only of *explaining* them (in such a way that they are not *res* or substances): Descartes has given a positive account of the "quality" that under different circumstances is called "heat" and "light" (AT 11:9). "Heat and other sensible qualities, insofar as they are in the objects, and also the forms of purely material things, as, e.g., the form of fire, arise from the local motion of certain bodies, and then themselves effect other local motions in other bodies." From this Descartes infers that these forms and qualities themselves consist in the size, shape, and local motion of the parts of the bodies: "[W]e have never noticed that what in external objects we indicate by the names of light, color, odor, flavor, sound, heat, cold, and other tactile qualities are anything other than the different dispositions [consisting in size, shape, and motion] of these objects that bring it about that they move our nerves in different ways" (*Principles* IV, 198). There is nothing anywhere in Descartes to suggest that he thinks that bodies are not really colored or hot and cold: "When we see some body, we have no more certainty that it exists insofar as it appears figured, than insofar as it appears colored; but we recognize much more evidently what *being figured* is in it than what *being colored* is" (*Principles* I, 69).

real quality is a quality that is a *res,* so a substantial form is a form that is a substance. Qualities too are forms, although "accidental forms," and a philosopher who thinks that all forms are accidents will reject substantial forms while continuing to believe in forms.[7] This was a common stance in the seventeenth century: Robert Boyle's *Origin of Forms and Qualities* rejects *substantial* forms and *real* qualities, and the Port-Royal *Logic,* which calls for eliminating "a certain bizarre kind of substances called in the School 'substantial forms,' " complains of Aristotle's *Physics* not that it is false, but that it is trivially true:

> For who can doubt that all things are composed of a certain matter and of a certain form of this matter? Who can doubt that, for this matter to acquire a new mode and a new form, it must not have had this beforehand, that is, that it must have had its privation? Finally, who can doubt these other metaphysical principles, that everything depends on the form, that matter by itself does nothing, that place, motion, qualities, and powers exist? But after we have learned all this, it doesn't seem that we have learned anything new, or that we are any better able to give an account of any of the effects of nature.[8]

7. Contra Gilson (1984, 162–63), Descartes is correctly following Scholastic usage in saying that a substantial form is a form that is a substance, while an accidental form is an accident (see, e.g., Suárez, disp. 15, sec. 1, pars. 5–6). Descartes makes his meaning clear in writing to Regius: "Lest there be any ambiguity in the word, let it be noted here that by the name 'substantial form,' when we deny it, is understood a certain substance adjoined to matter, and composing with the matter a merely corporeal whole, which [form], not less but even more than the matter, is a true substance, or *res* subsisting by itself, since indeed [the form] is said to be act, and [the matter] only potency" (AT 3:502). Descartes shows he knows that the Scholastics also recognized nonsubstantial forms when he says further that "all of the reasons [presented by Voetius] for proving substantial forms can be applied to the form of a clock, which no one will say is substantial" (AT 3:505; cf. 2:367). It misses the point to say that Descartes, in rejecting substantial forms and real qualities, is accepting qualities and forms but "not in the Scholastic sense"; he is making a precise anti-Scholastic statement within the Scholastic vocabulary, without twisting or deconstructing that vocabulary. Most discussions of Descartes' "rejection of hylomorphism" are impeded by a tendency to take "substantial" form and "real" quality and "prime" matter as pleonastic expressions for the Scholastic conceptions of form, quality, and matter. Even to speak of a "rejection of hylomorphism" is dubious: on their own self-understanding, seventeenth-century philosophers did not reject conceptual schemes (or whatever hylomorphism is); rather, they rejected principles, where principles are *things* (*substantial* forms, *real* qualities, *formless* matter). Better yet, they did not so much reject these principles as *abstain* from them as unclear and unnecessary.

8. Arnauld 1775–83, 41:122. Arnauld endorses forms, saying that "the form is what renders a thing thus-and-such, and distinguishes it from others, whether it is a being really distinguished from the matter, according to the opinion of the School, or whether it is only the arrangement of the parts; it is by the knowledge of this form that one must explain its

The question whether any forms are substantial is connected with the question whether any forms are real. For the Scholastics, some forms are substances, and some are accidents; some of these accidents are real, and some are merely modes, but every substance is a *res*. But Descartes maintains, against the Scholastics, that every *res* is a substance and therefore that all accidents are merely modal[9]—except inasmuch as a substance can be an accident of another substance, as a piece of clothing (which is a substance) belongs to a person as an accident in the category of habit.[10] Descartes' reasoning is simple: "[I]t is altogether contradictory for there

---

properties" (309); a few pages later he gives an all-out denunciation of "a certain bizarre kind of substances called in the School 'substantial forms'" (312), than which "nothing is less well founded" (313). Similarly, Boyle accepts forms (1991, 40, 52, 53–54, 62, 69, etc.), and qualities including sensible qualities (13–15, 28–37, etc.), all the while polemicizing against the doctrine of *substantial* forms and *real* qualities, which he, like Descartes, understands as the thesis that some forms are not accidents but substances (53–57) and that some qualities are not modes of matter but real beings really distinct from it (15–16, 22, 25, 28, 31). Malebranche too concludes the chapter of the *Recherche de la vérité* against substantial forms (bk. 1, chap. 16) by noting that "there is nothing to be said against these terms 'form' and 'essential difference'; doubtless honey is honey through its form, and it is by this that it differs essentially from salt, but this form or essential difference consists only in the different configuration of its parts." As far as I know, no seventeenth-century philosopher denied the existence of forms in bodies; the question was only whether some of these forms were substances.

9. The Scholastics also recognize what they call "substantial modes," such as the mode of union between a substantial form and its matter, which contribute to constituting a complete substance, rather than attaching to an already complete substance (Suárez, disp. 32, sec. 1, par. 15). These modes are not themselves properly substances, but belong to the category of substance improperly or "reductively." Suárez (disp. 32, sec. 1, pars. 13–19) discusses the relationship between the substance-accident and *res*-mode divisions; see Menn forthcoming. Since, for Descartes, the two divisions collapse, he often speaks of a substance-mode rather than *res*-mode or substance-accident opposition. But it is important to recognize that Descartes is not simply substituting a new metaphysical vocabulary for the old Scholastic vocabulary (as Daniel Garber suggests [1992, 68–70]); here, as with forms and qualities, he is making a precise anti-Scholastic statement in the Scholastic vocabulary. (Garber is rather careless with Scholastic terminology in this passage: it is not a Scholastic view, as he says, that "a substance is intimately linked with certain accidents, those that constitute its form or nature or essence" [68].)

10. In the Sixth Replies Descartes affirms once, and denies once, that a substance can be an accident of another substance. Descartes says that "clothing considered in itself is a substance, although it is a quality as referred to the clothed man; and the mind too, although it is really a substance, may nonetheless be called a quality of the body to which it is conjoined" (AT 7:441–42); but earlier he has insisted that although "one substance can belong [*accidere*] to another substance, still, when this happens, it is not the substance itself, but only the mode by which it *accidit* [i.e., its mode of belonging to or union with the other substance] that has the form of an accident; as, when clothing *accidit* to a man, not the clothing itself, but only the being clothed, is an accident" (AT 7:435). (In the first passage,

to be real accidents, since whatever is real can exist separately from every other subject; but whatever can exist separately in this way is a substance, not an accident" (AT 7:434). The Scholastics will object that substances are defined not by their ability to subsist by themselves but by their ability to receive contrary attributes; they will say that an accident, even when it does not *actually* inhere in anything, still has the *aptitude* to inhere; but this sounds forced, and there is some merit in Descartes' claim that the Scholastics are implicitly thinking of real accidents as little substances.[11] Descartes criticizes the Scholastics, not for positing qualities and forms that do not really exist, but for ascribing too high an ontological status to the qualities and forms that do exist: fire really is hot, but heat is just a mode, and the Scholastics give it too high a status when they say that it is a *res,* and when they think (implicitly) that it is a substance.[12] In criticizing this Scholastic error, Descartes' intention is not simply to deny the reality of sensible qualities but "to explode the reality of accidents" in general (AT 7:434), whether these accidents are merely sensible, like colors, or whether they are intelligible, like figure and motion.

On this account it becomes much easier to see why a philosopher would want to deny that sensible qualities are real, and why this denial would not draw protests from outraged common sense. But then a new

---

as often, Descartes is using *quality* broadly for *accident* in general.) There seems to be no real issue between the two ways of speaking. Suárez recognizes that clothing, as an accident in the category of habit, is a substance denominating another substance extrinsically (disp. 53, sec. 1; the Scotists, by contrast, speak as Descartes does at AT 7:435). Descartes insists that *all* real accidents must be analyzed analogously.

11. "When I conceived of heaviness (for example) after the kind of some real quality that was present in gross bodies, then although I called it a *quality* insofar as I referred it to the bodies in which it was present, nonetheless since I added that it was *real,* I was really *[revera]* thinking that it was a substance, just as clothing considered in itself is a substance, although it is a quality as referred to the clothed man; and the mind, too, although it is really a substance, may nonetheless be called a quality of the body to which it is conjoined" (AT 7:441–42). Similarly, Descartes tells Elizabeth that when "we" were involved in the errors of the Scholastics, we attributed certain notions "to the various qualities of bodies, as to weight, heat, and others, which we imagined to be *real,* that is, to have an existence distinct from that of the body, and in consequence to be substances, even though we named them 'qualities'" (AT 3:667). For the thesis that aptitudinal (rather than actual) inherence is essential to accidents, see Suárez, disp. 37, sec. 2; and that the ability to receive contraries is proper to substances (although the ability to subsist by themselves is not), see, e.g., Ockham, *Summa logicae* 1.43 (the sixth *proprium*), following Aristotle, *Categories* c.5.

12. "[W]e do not deny active qualities, we just deny that any more-than-modal entity should be attributed to them; for this cannot be, unless they are conceived as *[tanquam]* substances" (AT 3:503).

puzzle arises: if Descartes, in denying that heat is a real quality, does not deny that bodies are really hot, why does he deny that the heat in bodies "resembles" our idea of heat, or that the term *hot* can enter into scientific explanations? Certainly heat is not a *res,* but neither is sphericity, and this does not stop the term *sphere* from entering into scientific explanations. It is also puzzling why (as Descartes thinks) the prejudices of the senses should incline us to believe that heat is a real quality: do the senses really have a view on which qualities are *res* and which are modes? Again, if the senses do have views on such abstruse ontological questions, why does the practicing scientist have to go against them? To understand how the question of the ontological status of accidents is connected with Descartes' scientific program, we must examine more closely what it means to say that *x* is a *res* or that *x* is a mode: and we can see this best in the Scholastic discussions that engendered the terminology of *res* and mode.

For the Scholastics, the problem of the ontological status of accidents arises out of the analysis of predication. Whenever we make an assertion of the type "*a* is *b*," then (on the Scholastic realist analysis), the intellect conceives of the predicate-term "*b*" as signifying a *res,* the form *b*-ness, inhering in another *res,* the *suppositum a.*[13] But the Scholastics recognize many cases of nonstandard predications, where the proposition "*a* is *b*" can be true even though there is no *res b*-ness really distinct from *a* and really inhering in *a*. In an essential predication like "man is an animal," the predicate-term signifies a *res,* but a *res* really identical with the subject (*a* and its *b*-ness are said to be *rationally distinct* because, though really identical, they are represented by reason as if they were distinct). This cannot happen when the predication is accidental (if *a* could still exist without its *b*-ness, it cannot be identical with its *b*-ness); but it might happen that the predicate does not signify a *res* at all, as in "Socrates is blind," where there is no *res* blindness, or in "Socrates is known by Plato," where there is no *res* knownness. Blindness and knownness are said to be *entia rationis* (in the former case a privation, in the latter case a relation of reason), because, though they are not really beings, they are repre-

13. As St. Thomas says, "what the intellect puts on the subject side, it ascribes to the side of the *suppositum;* what it puts on the predicate side, it ascribes to the nature of a form existing in the *suppositum*" (*Summa theologiae* 1.q13a12). This is only the realist view; for simplicity, I will avoid discussing the nominalist theory of predication. Although Descartes rarely focuses his discussion on predication (or on any other topic in logic: he has a low opinion of the whole subject), it is clear from many incidental references that he presupposes a Scholastic realist theory of predication (by contrast, Hobbes, who was much more interested in logic, essentially repeats Ockham's account in the first part of his *De corpore;* Calvin Normore and I will compare Descartes and Hobbes on predication in our book *Nominalism and Realism*).

sented by reason as if they were beings.[14] (On another [compatible] analysis of "Socrates is known by Plato," the predicate signifies, not a knownness in Socrates, but a knowledge in Plato; here the predicate signifies a *res*, but a *res* not present in *a*, so *a* is said to be *b* by *extrinsic denomination*.) Now St. Thomas seems to think that if "*a* is *b*" is a true accidental predication, and if *b*-ness is not a negation or a relation of reason, and if *b*-ness is present in *a*, then *b*-ness must be a *res* really distinct from *a*. But for later Scholastic realists like Suárez, there is still another question to be asked: is *b*-ness a *res* really distinct from *a*, or is it merely a *mode* of *a*, a *way* a *is*, which is not itself another *res*?

For Suárez, this is equivalent to the question whether *a* and its *b*-ness are separable from each other at least by divine power. If *b* is accidental to *a*, then *a* can exist without its *b*-ness, but it is more obscure whether *b*-ness can also exist by itself, without inhering in its subject *a* (or in any other subject); if this is not possible, then *b*-ness is not a *res*, but merely a mode of *a*.[15] In general, a form is a mode of its subject (or "modally distinct from it *ex natura rei*") if the subject can exist without the form but the form cannot exist without the subject; whereas if *either* can exist without the other, they are really distinct as one *res* from another, and if *neither* can exist without the other, they are only rationally distinct.

14. Negations and relations of reason are the only kinds of *entia rationis*, since "every absolute [versus relative] positing [versus negation] signifies something existing in the nature of things" (Thomas, *De veritate* q21a1); some Scholastics distinguish privations as a third kind of *entia rationis*, while others include them under negations. There are also *entia rationis sine fundamento in re* (chimeras), but these arise only in false judgments; I am interested here in the ones that can arise in true (affirmative) judgments. All these quasi-beings are called *entia rationis*, not simply because they have their quasi-existence in relation to the intellect, but because the intellect apprehends them *as if they were beings*: "Since being is the primary object of the intellect . . . the intellect cannot know its opposite, namely nonbeing, except by somehow imagining [*fingendo*] it as being: and when the intellect tries to apprehend this, an *ens rationis* is produced" (Thomas[?], *De natura generis* c.1). See Suárez, disp. 54.

15. As Suárez notes (disp. 7, sec. 1, par. 19), *mode* is sometimes used more loosely, either for a *res* that modifies another *res* (as in the Thomist description of quality as *modus substantiae*) or for fundamental ways of being such as infinity and finitude, which are only rationally distinct from the *res* they modify. But for Suárez, modes in the strict sense are non-*res* that are modally distinct *ex natura rei* from the *res* they modify, where the test of a modal distinction is that the *res* can exist without the mode but not vice versa (see Menn forthcoming). Note that the principle that God can make any *res* without any other *res* breaks down for modes, not just in that modes cannot exist without *res*, but also in that (in a sense) *res* cannot exist without modes: God can create the *res* *x* without *any* real accidents, and he can create it without *any given* mode, but he cannot create it without *any* modes at all (e.g., he must give it either subsistence or inherence, and he must give any extension *some* figure).

Suárez thinks it is easy to show that some accidents are real and some are merely modes. Suárez does record the opinion of "the pagan philosophers," including Aristotle, that "the essential *ratio* of every accident consists in actual inherence in a subject," that is, that no accident can exist by itself without inhering in something else; and "this would be most true if, as seems to have been the opinion of the ancient philosophers, accidents are not *res* on their own, but are merely modes of a first subject." But as Suárez immediately insists, "the Catholic faith condemns this opinion, at least to the extent that it cannot be true universally," since in the Eucharist the accidents of the bread and wine continue to exist without inhering in any substance:[16] the pagan philosophers had never seen an accident existing by itself, but every Catholic philosopher has seen God conserve accidents without their subjects, and so he knows that at least some accidents are *res* really distinct from their subjects. But it is also easy to show that some accidents are merely modal. The clearest example is the mode of inherence: if whiteness inheres in a piece of bread, then inherence is in the whiteness; the inherence is accidental to the whiteness and cannot be really identical with the whiteness, since (if whiteness is a real quality) the whiteness can continue to exist without inhering; but the inherence cannot be a *res* really distinct from the whiteness and inhering in the whiteness. For if it were, this inherence could exist without inhering in the whiteness, and so the inherence would have its own inherence as a further accident; if every inherence were a *res* and thus separable from its subject, the white bread would contain an infinity of really distinct inherences. Clearly the regress must terminate in some inherence that is a modal accident and cannot exist without inhering in its subject (this inherence will be really identical with its own inherence,

16. Suárez, disp. 37, sec. 2, pars. 2–3. We know that the accidents of the bread and wine continue to exist, since we still see and otherwise sense them. But the accidents do not continue to exist in the substance of the bread and wine, since these substances are no longer present; nor do they inhere in the body of Christ, both because a glorified body is impassible (and so cannot receive new qualities) and because the body of Christ would have to receive contrary accidents (being leavened in Constantinople and unleavened in Rome, etc.). It is not necessary that *all* the accidents of the bread and wine subsist without a subject, though none of them have a *substantial* subject: some of the accidents may inhere in others, so that only the most basic accidents subsist without any subject at all. Indeed, on the Scholastic realist account, the qualities and other accidents of the bread and wine inhere in the quantity, and the quantity subsists by itself. On all this, see, e.g., St. Thomas, *Summa theologiae* 3.q77a1–2; the nominalists think that the qualities subsist by themselves, and that neither here nor elsewhere is there a quantity distinct from substances and qualities.

so it blocks the regress). The only reasonable conclusion is that the initial inherence of the whiteness in the bread is itself such a modal accident of the whiteness.[17]

It can be difficult to tell whether a given accident is a *res* or a mode. Suárez seems to grant a presumption in favor of reality: for God can always create an accident without its subject unless this would involve a logical contradiction, and even if we cannot clearly conceive how the accident could exist without a subject, we should assume that God can bring this about unless we clearly perceive that he cannot. Still, in many particular cases Suárez argues that some accident must be merely modal: this includes all the accidents in the categories of action, passion, where, when, and position, and in the category of quality, it also includes figures. If a body is now cylindrical, God can make it spherical simply by moving its parts locally, without also creating a real accident of sphericity; since God can make a body spherical without any *res* sphericity, it is superfluous to posit any such *res,* and we should conclude that the sphericity in a spherical body is not a *res* but simply a mode of the extension or continuous quantity of the body, a way of being extended that is not a *res* beside the extension. A figure is a "mode of termination," the way some extension ends: sphericity is being-extended-equally-far-from-the-center-in-every-direction; it is not some *res* added on to terminate the extension.[18]

The example of figure suggests a general rule for deciding whether a given quality-term signifies a *res* distinct from its subject:

> When predicables that cannot be *simultaneously* verified of the same thing can be *successively* verified of the same thing on account of local motion alone, then these predicables need not signify distinct *res.* "Curved," "straight," and the like are of this kind. . . . But this is not so with whiteness and blackness, since something does not become white or black, or hot or cold, merely because its parts are moved locally; and therefore all such things involve *res* distinct from the substance.

17. This is essentially Suárez' argument (disp. 7, sec. 1, pars. 17–18). For more references and a full discussion, see Menn forthcoming.

18. Figure or shape is what results from the outline of a single body (as Suárez puts it, "a mode resulting in a body from the termination of a magnitude" [disp. 42, sec. 4, par. 15]; Descartes describes it as *terminus rei extensae* [AT 10:418]); it is not equivalent to the modern concept of the "configuration" of a system, as the total state of the system resulting from the arrangement of all its parts. Thus the mechanists always say that the qualities of a body are determined by the "figure, size, and motion" of each of its parts, not simply by the figure of the whole body.

This quote is from Ockham (*Quodlibet* VII, 2; cp. *Summa logicae* I, 55), but it could equally have been from Suárez, and it gives the general rule that Suárez in fact uses for determining which qualities are real.[19]

From these texts, which are typical of the Scholastic discussion of real qualities, we can see why Arnauld immediately raised the example of the Eucharist: this was the standard (and, as Suárez thinks, decisive) objection to anyone who denies that qualities are real. Descartes had anticipated the objection as far back as 1630, when he told Mersenne that "wishing to describe colors in my way [in a *Dioptrics*], [I was] consequently obliged to explain how the whiteness of the bread remains in the holy Sacrament" (AT 1:179). Descartes thought he could deal with the problem, at the cost of rewriting many details of Scholastic theology, of struggling to fit various doctrinal pronouncements, and generally of getting in over his head. From the Ockham text, we can see why he thought it was worth it. If a quality of bodies, such as heat, is real, then no local motion of the parts of a body can be either necessary or sufficient for the body to become hot; but if heat is merely a mode of the extension of a body, then changes of heat, like changes of figure, can be explained entirely by local motion. It is thus crucial to Descartes' program in physics to deny the reality of heat: if Descartes is right that our sense organs are affected only by a mechanical communication of motion, then at least the *sensible* qualities of bodies (since these are the causes of our sensations) cannot be real:

> To explode the reality of accidents, I see no need to look for any other arguments than those I have already given. In the first place, since all sensation takes place by contact, nothing can be sensed beyond the surface of bodies; but if there are any real accidents, they must be something different from this surface, which is merely a mode [sc., the mode of termination of a body and its surrounding bodies]; therefore, if there are any [real accidents], they cannot be sensed. But who has ever thought that they existed, except because he thought that they were sensed? . . . But since the chief reason that has moved philosophers to posit real accidents has been that they thought sense perceptions could not

19. Suárez and Ockham agree that, because God can make a cylinder spherical just by changing how its parts are located (not creating any *res* sphericity, and annihilating one if it arises spontaneously), we should not posit that *spherical* signifies a *res* besides these parts; the difference is that Suárez thinks that *spherical* signifies *the way* the parts are located, and that this way is a mode *ex natura rei* distinct from the *res,* while Ockham refuses to admit such quasi-entities. (Another difference is that, for Suárez, the immediate subject of the figure is a continuous quantity rather than a corporeal substance, while Ockham refuses to draw such a distinction.)

be explained without them, I have promised to explain each of the senses in detail in my *Physics* [i.e., the *Principles*]; not that I want anything taken on faith, but from what I have already explained about vision in the *Dioptrics,* I thought that those who judge aright could easily conjecture what I could do in other cases. (AT 7:434–45)[20]

What Descartes is here promising in the *Principles* is just what he had earlier promised in *Le monde:* to show that by positing only God and minds and extended matter (moved and shaped in various ways), we can derive all phenomena, and all sensory perceptions, that we observe in the actual world. If Descartes can make good on this promise, then God *could* have created a world indistinguishable from the actual world without creating any *res* (except human minds) in addition to extended matter; and if so, we should believe that God did in fact create that world, and we should not posit any real accidents or substantial forms in nature.[21]

20. This passage of the Sixth Replies refers back to an argument of the Fourth Replies (AT 7:249, 250–51) that what in bodies immediately affects our senses is only the surface at which we touch the bodies, and that this surface is "not any part of the substance or of the quantity of that body, nor a part of the surrounding bodies," but only a mode, namely the mode of termination of the body, or the mode of union between it and the surrounding bodies. Here Descartes is taking a stand in a Scholastic debate about the status of surfaces as one of the species of quantity, namely two-dimensional continuous quantity (this includes Aristotelian place as "the surface of the surrounding body," as Descartes himself notes at AT 7:434 and 3:387); for this debate see, notably, Ockham, *Summa logicae* 1.44–46 and Suárez, disp. 41, secs. 5–7. Descartes apparently thinks (AT 7:433) that the consensus view of "mathematicians and philosophers" is that surfaces are merely modes, *if* surfaces are understood strictly, not just as thin bodies whose depth is ignored, but as lacking depth altogether, so that two contiguous bodies have precisely the same common surface. Ockham insists that surfaces are only bodies with depth ignored, and Descartes thinks that this is an error we are naturally led into by our imaginative representation of a surface (discussion in Rule 14, in AT 11:445–49); since, Descartes says, even geometers frequently fall victim to this, this shows that even a geometrical representation can deceive us by representing a non-*res* (a mode) as if it were a *res*. In fact, Suárez thinks surfaces are *res* distinct from three-dimensional quantities, but it would have been more consistent with his general program, in answering Ockham's arguments, to interpret them as modes of termination and union, as Descartes thinks the Scholastics generally did and as many of them doubtless did.

21. This is the strategy throughout *Le monde*, made clearest in the "fable" of a "new world" (AT 11:31ff.). Descartes' argument against positing a form of fire, a quality of heat, and an action of burning really distinct from the particles of the wood, by the thought-experiments of annihilating the forms while leaving all the motions of the particles, or keeping the forms while stopping the motions (*Le monde,* in AT 11:7–8), is very close in spirit and execution to the Scholastic voluntarist arguments by which Ockham and Suárez conclude that figures, actions, and so on are not real accidents. A comparison leaves no doubt that Descartes was aware of such Scholastic arguments and wanted to extend them to prove much more radical conclusions.

Indeed, Descartes proposes to reduce the physical world, not merely to extended matter, but to extension or continuous quantity alone: he argues that there is no need to posit a matter really distinct from this quantity (*Principles* II, 8–9). For Scholastic realists, continuous quantity holds a privileged position among accidents, mediating between substance and the other accidents. Although qualities are not qualified, quantities are quantified: the coldness of a stone is not cold, but its one-foot-long-ness is one foot long, and in general, every continuous quantity is coextensive with the substance it quantifies. Qualities are also coextensive with their substances, but this is only because the qualities are quantified, and this is because the qualities proximately inhere in the quantity, which in turn inheres in the substance: indeed, continuous quantity is a quasi-substance extended throughout the physical world, which is the immediate substrate for all other accidents.[22] As St. Thomas notes, continuous quantity can be conceived by the intellect separately from all substance, and is so conceived in geometry; it follows, since "God can do more in actual production than the intellect can in apprehension," that God can create continu-

---

22. Descartes was aware of this Scholastic position and apparently regarded it as the normal background assumption: "[T]here is no incompatibility or absurdity in saying that one accident is the subject of another accident, as it is said [*on dit*] that quantity is the subject of the other accidents" (AT 3:355). Ockham gives a brief statement of the realist doctrine of continuous quantity, along with his arguments against it, in *Summa logicae* 1.44 (Ockham thinks that quantities, whether continuous or discrete, are nothing beyond the substances or qualities that they quantify). Suárez presents one realist theory of continuous quantity (and discusses other opinions) in Disputation 40: in section 1, paragraph 6, he notes the "peculiar condition of quantity, which is not only the form by which something else is *quantum*, but is also itself denominated *quanta*, since it is not only the *ratio* on account of which other things become extended and divisible, but is also extended and divisible in itself; nor could it extend something else unless at the same time it were coextended with it, and had its own parts corresponding to the parts of its object." This self-predication makes quantity more substancelike than the other accidents, so that "it does not have its essential *ratio* in relation to substance, but in relation to itself; whence in the mathematical sciences it is considered abstractly, as if it existed by itself without any relation to a substance" (disp. 37, sec. 2, par. 3). In Disputation 14, section 4, Suárez endorses the usual realist view that quantities are the proximate subjects of corporeal qualities ordinarily, and their ultimate subjects in the eucharistic species. Although it is agreed that the essence of continuous quantity is extension, it is controversial how this extension should be interpreted: Suárez thinks it is neither the distinctness of the parts of the substance, nor their *actual* extension or size, but their *aptitudinal* extension, that is, their *tendency* to occupy a determinate amount of space, and to resist being compressed further or becoming coextended with each other; so a substance can be rarefied or condensed while keeping the same quantity (disp. 40, sec. 4). This is apparently the view Descartes mocks in *Principles* 2.5, of "some who are so subtle that they distinguish the substance of a body from its quantity, and then distinguish this quantity from extension" to account for condensation and rarefaction.

ous quantity without any substance distinct from that quantity, thus actualizing something like Plato's separate mathematicals.[23]

This is just Descartes' argument. In the Fifth Meditation Descartes claims a distinct perception of "that quantity which the philosophers commonly call 'continuous,' or the extension of that quantity (or rather of the quantified thing) in length, breadth, and depth" (AT 7:63). This quantity or extension, which is itself extended, is the object of geometry, and, Descartes argues in the Sixth Meditation, God can actualize it by itself. "I know that [material things] are able to exist inasmuch as they are the object of pure mathematics, since I perceive them clearly and distinctly: for there is no doubt that God is capable of producing all things that I am capable of perceiving in this way" (AT 7:71). And "since I know that all things that I clearly and distinctly conceive can be produced by God in the way that I understand them, it is enough that I should be able to understand one thing clearly and distinctly without another, in order to be sure that one is diverse from the other, since it can be produced separately, at least by God" (AT 7:78). So God can create continuous quantity without also creating any matter or forms or qualities really distinct from it. The Sixth Meditation argues that God has created continuous quantity outside us, and if we can explain the phenomena without any *res* beside this, we should believe that this is *all* that God has created. Descartes' Scholastic opponents, like himself, distinctly conceive this *res* continuous quantity, and they recognize its existence in bodies; but then, dissatisfied with what they distinctly understand, they suppose that this is merely an accident of some confusedly imagined substance. Descartes proposes to explain everything through what everyone clearly understands, quantity, without positing any substance or qualities really distinct from it.[24] This, then, is how Arnauld knew that "Descartes thinks there are no sensible qualities, but only the various motions of the cor-

---

23. The comparison with the Platonic mathematicals is Thomas' own. Thomas Aquinas, *Summa contra Gentiles* 4.65. Thomas thinks that in the eucharistic species God does in fact conserve quantity without substance, although not without qualities inhering in the quantity.

24. In *Principles* 2.8 Descartes argues that "quantity does not differ from extended substance *in re,* but only in our conception, as number does from the thing numbered." This is in agreement with Ockham, but where Ockham had used this claim to eliminate the superfluous realist quantities in favor of the commonly accepted *res quantae,* substances and qualities, Descartes uses the same claim to reduce the *res quantae* to their intelligible quantities, eliminating any obscure subject distinct from the quantity itself (Descartes had taken this route as early as Rule 14, which already showed familiarity with Scholastic debates about quantity). In *Principles* 2.9 Descartes diagnoses the error of his Scholastic opponents: "[A]lthough some people may *say* otherwise, I don't think that they *perceive* otherwise about this matter; but when they distinguish the substance from the extension or quantity,

puscles touching us, by which we perceive those various impressions that we then call color, taste, and smell; so that there remain [only] figure, extension, and mobility," which are only modally (or, as Arnauld says, "formally") distinct from bodily substance (AT 7:217–8). Arnauld has simply noticed that Descartes has not argued or assumed that God has created any *res* in bodies beyond extension; so Arnauld assumes, correctly, that Descartes believes that there is nothing in bodies beyond the bare minimum, extension and its modes.[25]

We can now clear up some of the puzzles that were raised earlier. We have already seen why Descartes thought it was important for the physicist to recognize that the qualities of bodies are not real; we can now see why Descartes thinks that sensible-quality terms like *hot,* and other terms signifying active and passive powers in bodies, should not be admitted in scientific explanation. The term *hot* should not be admitted, at least not as long as it remains the expression of our sensory idea of heat, *not* because heat is not real (since figures are equally unreal), but because the sensory idea of heat is confused and does not represent heat as it is. The

---

either they understand nothing by the name 'substance,' or they have only a confused idea of incorporeal substance, which they falsely ascribe to corporeal substance; and they consign the true idea of this corporeal substance to 'extension,' which they call an accident, and so they express in words something quite different from what they comprehend in their minds." It is important for Descartes that his opponents have the true distinct idea of corporeal substance (though they refuse to *call* it corporeal substance), so he can claim that he is not introducing any new principles beyond those that everyone understands and accepts (he merely abstains from some old principles). Seeing this allows us to solve a problem that Garber raises for Descartes' argument that the essence of body is extension: "What Descartes needs to establish is that our idea of body is the idea of a thing whose *only* properties are geometrical, a thing that *excludes* all other properties [including all sensory qualities]. But what emerges from the argument from elimination is the idea of a body as a thing at least *some* of whose properties are required to be geometrical. . . . from the fact that *some* bodies are not colored it does not follow that *no* body is really colored, any more than it follows from the fact that some bodies are not spherical [so that sphericity is not 'essential' to body] that no body is really spherical" (1992, 80). But Descartes' argument is safe and (among realists) uncontroversial in isolating a *res,* quantity, which can exist by itself and contains nothing distinct from extension and its modes; Descartes' real burden will be, not to show that this *res* has no other properties (and not to show, what Descartes does not believe, that it is not really colored), but to persuade us that God has created no other *res* attached to this (except human minds), and therefore that this alone deserves the title "body."

25. Arnauld may also have in mind *Dioptrics* (AT 6:84–85), suggesting that light is just a pressure transmitted by the air, and that colors are just the different ways in which bodies receive and reflect this pressure; but this contains nothing nearly as clear, or as radical, as a reduction of all qualities of bodies to modes of extension.

sensory idea of heat does not represent heat as it is, because although heat is only a mode, our idea does not display heat as a mode but instead represents heat confusedly as if it were a *res*.

Of course (since heat is not in fact a *res*) we cannot *distinctly* perceive that heat is a *res:* there is no contradiction in asserting that "heat is [not a *res* but] simply the agitation of particles of the third element" (cf. *Principles* IV, 29). Nonetheless, Descartes thinks that our sensory idea of heat represents heat as a *res* or as if it were a *res (tanquam rem),* and that, if heat is not in fact a *res,* the idea will be "materially false" (AT 7:43): that is, without being properly or formally false (since falsehood properly belongs only to judgments), and without necessarily falsifying every judgment in which it occurs, the idea *gives occasion* for error, just by representing *non rem tanquam rem.*[26] It is not immediately obvious what it means for an idea to represent its object *tanquam rem.* Sometimes Descartes speaks as if every idea, just by being an idea and a representation, necessarily represented its object *tanquam rem,* so that every idea whose object is not a *res* would be materially false: "since there can be no ideas that are not *tanquam rerum,* then if it is true that cold is nothing other than the privation of heat, the idea that represents this to me as something real and positive *[tanquam reale quid et positivum]* would not undeservedly be called false; and so in the other cases" (AT 7:44). But this is an exaggeration, and there are legitimate ways of representing privations and other *non res:* Descartes says in the Fourth Meditation that he has "not only a real and positive idea of God . . . but also, so to speak, a negative idea of nothing" (AT 7:54), and there is no suggestion that this idea of nothing involves a mistake. But (as Descartes says when Burman notes the conflict between the two passages) "this idea is only negative, and can scarcely be called an idea; whereas [in the Third Meditation passage] the author was taking 'idea' properly and strictly" (AT 5:153). The point is that the idea of nothing does not simply represent a negative content but is itself the negation of an idea, and not an idea *simpliciter;* whereas an idea *simpliciter,* which does not manifest the negativity or

26. When Descartes introduces the notion of material falsity in the Third Meditation (AT 7:43–44), it is apparently definitional that ideas are materially false "if they are *non rerum*" or "when they represent *non rem tanquam rem*"; but in the Fourth Replies, "that some ideas are materially false" is "as I interpret it, that they provide the judgment with material for error" (AT 7:231). This latter criterion was not explicitly mentioned in the Third Meditation, but it gives the nominal sense of the phrase "materially false"; the Fourth Replies still insist that the real definition of material falsity, or the *reason* why an idea satisfies the nominal definition, is that it represents *non rem* (whether an objectively grounded negation or mode, or a mere fiction like a chimera) *tanquam rem.*

otherwise nonreality of its object, is *tanquam rei* and will be materially false if its object is not in fact a *res*.[27]

The most obvious way for an idea to be materially false is for it to represent a negation or privation without being itself a negative idea. This is the example Descartes uses in the Third Meditation to argue that our sensory ideas of heat and cold are confused: "[T]he ideas that I have of heat and cold are so little clear and distinct that I cannot learn from them whether cold is only the privation of heat, or whether heat is the privation of cold, or whether both are real qualities, or neither. And since there can be no ideas that are not *tanquam rerum*, then if it is true that cold is nothing other than the privation of heat, the idea that represents this to me as something real and positive *[tanquam reale quid et positivum]* would not undeservedly be called false" (AT 7:43–4). *Reale quid* here implies *positivum*: since a privation is only an *ens rationis* and not a *res*, if cold is a privation it fails to be a real quality.[28] But this is not the only way for cold to fail to be a real quality: since heat and cold might both

27. One might question what it means for such an idea to have a *non res* (say, a privation) as its "object" (compare Wilson 1990). Arnauld objects that, since the idea of *x* is just (the form of) *x* "itself, insofar as it exists objectively in the intellect" (AT 7:206), the idea of a privation must itself have privative form, and so will not be deceptive: "[T]his idea of cold, which you say is materially false, what does it exhibit to your mind? A privation? Then it is true. A positive being? Then it is not the idea of cold" (207). Descartes in his reply agrees that our positive idea of cold is not properly an *idea of* cold, since it is not (the privation) cold itself objectively existing in our mind, but "it often happens in obscure and confused ideas, among which these ideas of heat and cold should be counted, that they are referred to something other than what they are really ideas of" (AT 7:233). This sensory idea is not an *idea of* anything either positive or negative; that is, it is neither an external *res* nor the lack of one objectively present in my mind, but only a conventional sign, with no intrinsic objective content ("having no *esse* outside the intellect"), which God has arbitrarily established in my mind to signify the cold in bodies. But this idea deceives when it is "referred to" or "taken for" cold, although the idea does "represent" or signify the cold in bodies (and causes the associated word *cold* to denominate cold bodies), because it suggests that what it signifies is something positive, as it itself is: "I cannot discern whether it exhibits something to me that is positive outside my sensation or not; and therefore I have an occasion to judge that there is something positive, although perhaps there is only a privation" (233–34). The erroneous judgment arises when I take the idea to resemble *this* external thing: I am able to single out *this* thing in the judgment because it is what the idea is the *sign* of, that is, the condition in bodies that typically accompanies this idea, because God has established nature in such a way that this bodily condition *causes* this idea in my mind.

28. Wilson, apparently not recognizing the Scholastic contrast between *res* on the one hand and *entia rationis* (and modes) on the other, says that "Descartes *should* [though he does not] allow that the content of a distinct idea can be a privation, but not a non-thing" (1990, 19n6). Wilson wants to reduce the ontological question "Is *x* a *res*?" to the epistemological question "Can *x* be conceived distinctly?"; but figures and motions, as well as privations, can be conceived distinctly, and Descartes is emphatic that none of these are *res*.

fail to be real qualities, and since they cannot both be privations of each other, they would have to be *non res* in some other way. Descartes' point is that both heat and cold may be modes; since our ideas of these (unlike our ideas of figures) do not represent their objects as modes, they represent *non res tanquam res* and are thus materially false.[29]

This is supposed to explain why (as Descartes thinks) the prejudices of the senses incline us to believe, wrongly, that heat and cold are real qualities. But from a twentieth-century perspective, the *explanans* is more mysterious than the *explanandum*. Surely I can say, "Fire is hot, ice is cold," not only without *judging* that heat and cold are *res* (and not privations or modes) but also without even *suggesting* this, or giving occasion for a false judgment. But from the Scholastic realist perspective, what Descartes is saying makes perfect sense. It could have come straight from St. Thomas.

Thomas thinks that whenever the intellect forms a subject-predicate judgment, it conceives of the predicate as signifying one *res* inhering in another *res*: as Thomas says, "what the intellect puts on the subject-side, it ascribes to the side of the *suppositum;* what it puts on the predicate-side, it ascribes to the nature of a form existing in the *suppositum*" (*Summa theologiae* 1.q 13a12). But it may turn out that the predicate does not in fact signify any *res* inhering in the *suppositum;* indeed, in the passage I have just cited, Thomas is talking about what happens when we predicate something of itself, so that there is no composition of form and *suppositum* corresponding to our judgment. If the predicate does not signify a *res* inhering in the *suppositum,* then, as Thomas says, we are understanding the thing *aliter quam sit,* "otherwise than it is" (q13a12 ad3).

Not all judgments of this kind are false, however: as we have seen, they can be true when the predicate signifies a negation or privation or *relatio rationis,* or when what it signifies is really identical with the *suppositum,* or when it denominates extrinsically (or, for post-Thomist realists, when it signifies a *mode* of the *suppositum*). So Thomas must face the obvious objection: "[E]very understanding that understands a thing otherwise than it is, is false." (q13a12 obj3); so how can these other kinds of judg-

29. Alternatively, heat and cold might denominate extrinsically: they would then be *res* that are not *in* the hot and cold bodies, but denominate them hot and cold by some relation other than presence. In this case our ideas of hot and cold would be deceptive, not precisely by representing *non rem tanquam rem,* but by representing *rem non in re tanquam rem in re.* Although Descartes usually treats heat and cold as modes, at least one text apparently treats properly *sensible* qualities (but not active powers) as extrinsic denominations from our sensations. See n. 35 below.

ments be true?[30] Thomas replies that there are two ways of taking the phrase *aliter quam sit:* "[T]he adverb *otherwise* can determine the verb *understand* either on the side of the understanding or on the side of the thing understood." If it is taken the second way, Thomas grants the proposition that "every understanding that understands a thing otherwise than it is, is false"; this is equivalent to saying that "every understanding that understands a thing *to be* otherwise than it is, is false" (q13a12 ad3). But if the adverb is taken on the side of the understanding, Thomas denies the proposition, "for the understanding's way of understanding is not the same as the thing's way of being" (to say that the "ways" *[modi]* of understanding and being differ is to say that we understand the thing "otherwise" than the thing is). On Thomas' theory, we signify things, not necessarily in the way that they are in themselves, but in the way that we understand them (*modus significandi sequitur modum intelligendi;* q45a2 ad2). Typically the *modus intelligendi* in turn corresponds to the *modus essendi,* but sometimes the *modus intelligendi* will diverge from the *modus essendi,* and yet we can still signify the things and form true judgments about them.[31] In particular, whenever we form an affirmative judgment about God, we understand him "compositely" and thus under-

30. Note that Thomas and other Scholastics treat judgment as an act of the understanding *(intellectus).* Descartes argues in the Fourth Meditation that judgment is, rather, the will's assent to an idea presented to it by the understanding. This does not make too much difference for the issues I am concerned with here. The immediate question in St. Thomas is whether there can be true affirmative judgments about God: since there is no composition of any kind in God, every affirmative judgment about God must understand him otherwise than he is.

31. Thomas generally thinks that the form of our words reflects the form of our thoughts accurately enough, but that the form of our (true) thoughts may not reflect the form of the *res;* so we can name God in the way that we can understand him, but not in the way that he is in himself. In particular, a concrete name ("wise") signifies God as if he were a form-*suppositum* composite, and an abstract name ("wisdom") signifies him as if he were a form inhering in a *suppositum;* both are applicable to God, who is a simple subsisting form, but neither signifies him as he is (q13a1 ad2). The doctrine that Peter Geach (1972, 318–19) denounces as a "muddle" of the "scholastic manuals" ("that a thought of things *as being, as if they were,* what they are not, may both be inescapable for minds like ours and *not* be false thought") is in fact Thomas' own position. (Geach wants Thomas simply to be saying that "our mind in thinking need not . . . mirror the structure of the world," but not that it *normally* does, or that the discrepancy, when it does not, involves representing something *as if it were* what it is not, and so gives occasion for error.) Nor is the position as confused as Geach suggests: I can represent something in a way that involves a fiction, without assenting to that fiction. In a striking passage, Thomas actually says that, because of the divergence between the *modus intelligendi* and the *modus essendi,* we can legitimately *deny* such propositions as "God is wise" (though we can also, of course, legitimately affirm them): "[A]s far as the *res significata,* whatever is *aliquo modo* in God is truly attributed to him

stand him otherwise than he is; but our understanding is not false, since "it does not say *that he is composite,* but that he is simple" (q13a12 ad3).

The upshot is, for St. Thomas, that although there are true judgments that understand something otherwise than it is, these are true in an abnormal way. In the normal true judgment "Socrates is wise," the composition of the judgment reflects a structure *in re,* since wisdom is a *res* existing in Socrates; the abnormal true judgment "God is wise" fails to reflect the structure *in re,* and this is because, "on account of our intellect's connaturality to composite things . . . we can apprehend and signify simple subsistents only in the manner of *[per modum]* composite things" (q13a1 ad3, rearranged). Although both judgments correspond to reality in a weak, Tarskian sense of *correspond,* only the former judgment corresponds to reality in the stronger sense that it *structurally* corresponds.[32] But whenever we form a judgment, we are *tempted* to believe that our judgment structurally corresponds to reality, that we understand the thing as it is; so the true judgment "God is wise" tempts us into the false judgment "there is a real wisdom, really distinct from God and really existing in God." Much of the *Summa theologiae* (and of Suárez' *Metaphysical Disputations*) is devoted to refuting these falsehoods suggested by our true judgments; but even when we know that God is simple, it still *looks* as if there were a real accident of wisdom in God, just as the sun continues to *look* smaller than the earth. Where the predicate of a judgment signifies an *ens rationis,* it is fair to use Descartes' language and say that the concept of the predicate is "materially false," since it gives occasion for error and does so just by representing *non rem tanquam rem.*[33] In at least some

---

. . . but as far as the *modus* that they signify of God, they may be denied: for each of these names signifies some definite form, and in this way they are not attributed to God. . . . And therefore they can be denied of God absolutely, since they do not apply to him through the *modus* that is signified: for the *modus* that is signified is as they are in our understanding . . . but they apply to God in a higher *modus*" (*De potentia* q7a5 ad2). The denial is thus a legitimate (though extraordinary) precaution against errors I might be led into by the original true affirmation.

32. If I may be allowed an ethnic joke, you don't have to correspond to reality to correspond to reality, but it helps.

33. Similarly, when (as, for Thomas, in "God is wise") the predicate signifies something really identical with the *suppositum,* the occasion for error arises because we signify *unam rem tanquam duas res.* In this case we might prefer to say that the judgment itself (or, for Descartes, the composite idea to which the judgment is an assent) is materially false, although the judgment is formally true. Recall from the *De natura generis* that *entia rationis* are imagined *(ficta)* by the intellect as if they were real beings, and so give occasion for error. Seventeenth-century mechanists, picking up this Scholastic theme, also warn against this kind of temptation to error and use it to symbolize false positings in general: "[B]ecause we

cases, we can overcome the temptation to error by rephrasing the judgment in a form that does correspond to reality, as we can replace "Socrates is known by Plato" by "Plato knows Socrates."

Once we understand how the issue of the truth of our judgments is distinguished from, but also connected with, the issue of their (structural) correspondence to reality, we can see why Descartes thinks that we are tempted to believe that heat is a real quality in fire. Since we habitually make the true judgment that fire is hot, we are also tempted to make the false judgment that this first judgment corresponds structurally to reality, or that the ideas involved in this judgment represent things as they are: that is, that heat is a *res* really distinct from the fire and really present in the fire. Since heat is not in fact a *res*, we do not perceive heat as it is; this is what it means to say that heat *as we perceive it* is not in the fire. This is also what it means to say that the heat in the fire does not *resemble* our idea of heat: the resemblance we are tempted to believe in is a *structural* resemblance or correspondence between our judgment "Fire is hot" and the realities that make that judgment true.[34] Since heat is not a *res* and since the idea of heat is *tanquam rei*, the idea of heat gives occasion for error, and for this reason it should be avoided in scientific judgments.

---

have been conversant with them [sensible qualities] before we had the use of reason, and the mind of man is prone to conceive almost everything (nay, even privations, as blindness, death, &c.) under the notion of a true entity or substance, as itself [the mind] is, we have been from our infancy apt to imagine that these sensible qualities are real beings in the objects they denominate" (Boyle 1991, 31).

34. Twentieth-century scholars have caused much mischief by using such sentences as "Heat *as we perceive it* is not in the objects" or "The heat that is in bodies *does not resemble* our idea of heat," without inquiring into the meaning of the "as" phrase, or the sense in which ideas might be expected to resemble external objects; we can interpret these Cartesian affirmations only by understanding Descartes' general theory of cognitive representation, much of which is taken over from Scholastic realism. By interpreting them, instead, through vague common-sense notions (or through more recent philosophy), many scholars have concluded that Descartes denies that bodies are really colored, or that he can affirm this only by using a perverse sense of "color." Margaret Wilson writes, "I don't see that there can any longer be reasonable doubt that major early modern philosophers—with the exception of Berkeley—saw their commitments to mechanistic science as dictating acceptance of what has come to be called the 'error theory' with respect to colors, odors, tastes, sounds and the like: in seventeenth century terms, the claim that the senses deceive us in leading us to construe such experienced qualities as resembling real features of external objects" (1992, 234). The "error theory" is therefore supposed to be a translation of this seventeenth-century claim. Unfortunately, Wilson does not spell out what she means by "error theory," and different contemporary philosophers seem to use the term in stronger and weaker senses. Most strictly, it should mean that all (affirmative) color judgments are false. *Principles* I, 69–70 (etc.) makes it clear that Descartes did not believe this; nor did he believe the weaker claim that all color judgments *presuppose* some false judgment about the nature

Even though the idea of heat is materially false, it can still be used in true judgments; but in science we should be concerned not only about the content but also about the form of our judgments, not only that they are true but also that they represent things as they are, and so do not give occasion for error. So, Descartes says, "it is the same in content *[in re]*, if we say that we perceive colors in objects, as if we said that we perceive something in objects, we do not know what it is, but that produces in us a certain very clear and manifest sensation, called the sensation of colors. But there is a very great difference in the manner of judgment *[in modo judicandi]*" (*Principles* I, 70). Because of this difference in *modus judicandi*, we should beware lest the true judgment "Fire is hot" tempt us to believe that the *modus essendi* of heat in bodies is the same as our *modus judicandi*, and so tempt us to believe that we know what sort of thing heat is (at least, that it is a *res*), when in fact our senses tell us only what things are hot, and not what heat itself is. Once Cartesian physics has discovered what it is in bodies that causes the sensation of heat, then there is no objection to using the word *heat* to express the new distinct idea we will have of heat as a certain mode of extension.[35]

One major task of Cartesian physics is to explain what structures *in*

---

of colors. Descartes did think our color judgments structurally *suggest* (and are often accompanied by) a false judgment, and perhaps we could call this view an error theory. But if so, we should say what the error is: and it is not enough to describe it as the erroneous belief that colors in bodies *resemble* our ideas of color, unless we specify *in what respect* they are thought to resemble them. Wilson's statement of the theory in seventeenth-century terms is also unclear: apart from the difficulty about resemblance, "such experienced qualities" might mean either "our experience of such qualities" or "such qualities *as we experience them*," and, if the latter, the force of the "as" phrase is unclear; and "real features" is ambiguous between "features that objects really have" and "real accidents" in Descartes' sense.

35. This is how Descartes speaks of heat and other sensible qualities in the *Principles* and *Le monde*. But Arnauld attributes to Descartes a different doctrine of sensible qualities, according to which (although bodies are really colored, and although there are no real colors in bodies) colors are not modes but extrinsic denominations, denominating bodies from the sensations they cause in us; and there is at least one passage in Descartes that supports Arnauld's reading. "As for these Cartesians who are not willing to admit that our soul is green or yellow or stinking, I don't know what he [Malebranche, in the eleventh éclaircissement of the *Recherche de la vérité*] means. For if those he is speaking of claim that sensible qualities are modifications of extension, and not of our soul, they are not Cartesians on this point; but if, admitting that these are modifications of our soul and not of extension, they only maintain that this does not have the result that our soul should be called green or yellow or stinking, this is only a question of words, on which I don't believe they would be as wrong as this author imagines. We simply need to understand what is in question. Two Cartesians are going for a walk. One says, 'Do you know why snow is white, why coal is black, and why rotting carcasses smell so bad?' 'What silly questions,' answers the other. 'Snow isn't white, nor is coal black, nor do carcasses stink; it's your soul that's white when

rebus make our ordinary judgments true. Descartes is asking a very Scholastic question, but his conclusion is that the *modus essendi* of physical things differs from our *modus judicandi* much more radically than any Scholastic had believed. Descartes' treatment of the attributes of body is rather like the Scholastic treatment of the attributes of God: since God has given us the Scriptures, everything the Scriptures say about God must be true; but since the Scriptures, given to guide us in our weakness, are written in human language and suited to human modes of understanding, they do not represent God as he is (and so tempt us into false judgments). Theology has the task of explaining what structures *in rebus* make the

---

you look at snow, and black when you look at coal, and stinks when you're near a carcass.' I assume they agree on the basic doctrine, but I ask which of them has the better way of speaking? I maintain that it's the first, and that the other's criticism is unreasonable. For, to begin with, there are infinitely many *denominations* [Arnauld's italics mark the Scholastic technical term] that do not presuppose modifications in the things to which they are attributed. Is is speaking wrongly to say that the statue of Diana was worshiped by the Ephesians? But the honor these idolaters paid to the statue was not a modification of the statue, but only of the idolaters" (*Des vraies et des fausses idées*, in Arnauld 1775–83, 38:313). (Arnauld then lists further reasons why the first Cartesian's language is preferable, as corresponding both to God's intentions in giving us sensations, and to human intentions in giving meaning to sensible-quality terms.) The view that sensible-quality terms are said of bodies truly but by extrinsic denomination from human beings (like *healthy*, said of a food rather than of an animal) is supported by a passage of the Sixth Replies, in which the meditative persona, sorting through his ideas, "recognizes that nothing belongs to the essence of body, except that it is a long, broad, and deep thing, capable of various figures and of various motions; and that its figures and motions are just modes, which cannot exist without it by any power; but that colors, smells, tastes, and the like are only sensations existing in my thought, which differ from bodies no less than pain differs from the figure and motion of the projectile that induces the pain; and, finally, that heaviness, hardness, and the powers of heating, attracting, and purging, and all the other qualities that we experience in bodies, consist only in motion or the privation of motion, and in the configuration and location of the parts" (AT 7:440). This passage is curious in that it requires a sharp division between sensible qualities (analyzed as extrinsic denominations) and active powers (analyzed as modes); and yet several examples, notably heat and heaviness, seem to belong equally to both classes. Descartes might, like some Scholastics (cp. Suárez, disp. 42, sec. 4), distinguish two qualities of heat; but only shortly before (AT 7:434) he argued that sensible qualities cannot be distinct from surfaces because only surfaces *act* immediately on our senses. The truth is that Descartes has not worked out a consistent way of speaking and that he does not much care. Fire is really hot, and there is no real heat in fire, but only a mode in it that causes our sensation of heat; if we say that heat proper is our sensation, denominating the fire extrinsically, then the quality of heat must have a *fundamentum* that is a mode in the fire and a *complementum* in us. "Heat as it is in the body" (as Descartes sometimes says) is the mode, and usually Descartes is content to call it heat without qualification; but to remind us of its unlikeness to the heat that is in our minds, he is prepared on occasion to deny that it is heat, as Thomas is prepared to deny that wisdom as it exists in God is wisdom.

scriptural assertions true; and theology will in fact explain how these assertions can all be true of God even though God is entirely simple and incomposite. Likewise, for Descartes, God has given us nature as a guide, so what nature tells us about bodies must be true; but since nature speaks to us in the language of sensation, suited for practical guidance rather than for theoretical understanding, the teachings of nature do not represent bodies as they are (and so tempt us into false judgments), and physics has the task of explaining what structures in bodies make the teachings of nature true. Descartes is even ready to say that bodies are *simple* beings, (*Notae in programma,* in AT 8b:350–51), because, although they have a structure of parts, they do not have an inherence structure of *res in re.* For Descartes' Scholastic realist opponents, bodies contain first prime matter, then a substantial form inhering in the matter, then a continuous quantity inhering in the substantial composite, then real qualities (and whatever other real accidents there may be) inhering in the quantity. Descartes systematically eliminates all this composition of *res in re:* matter is really identical with continuous quantity (*Principles* II, 8–9, Rule 14, *Le monde,* in AT 11:35–36), forms other than the mind are not substances but simply collections of mutually sustaining qualities (AT 3:461, *Le monde,* in AT 11:26), and these qualities themselves are not *res* but simply modes of continuous quantity. By arguing that continuous quantity alone, without any additional matter or form or qualities, can produce the phenomena of the world we perceive, Descartes showed that the structure of the world can be radically different from the structure of our ordinary judgments about the world, so different that it becomes hopeless to investigate the world as the Scholastics did, by beginning with the structure of language and then noting the points at which the world diverges from our language. The structure of form in *suppositum* as *res in re* would be a linguistic structure in reality: but reality need not exhibit a linguistic structure at all. That, more than any particular reductionist program, and much more than the supposed doctrine that bodies are not really colored, is the lesson of Descartes' denial of real qualities.

ROGER ARIEW

༂13 ༈

PIERRE BOURDIN AND THE
SEVENTH OBJECTIONS

Bourdin's objections are probably the least successful of the sets of objections to Descartes' *Meditations*. They are probably also the least read and least appreciated. Part of the reason for that state of affairs is that they are the longest set of objections, even though they cover the least philosophical ground: the verbose Seventh Objections together with Descartes' even more verbose replies, mocking Bourdin's style, actually fill more pages than any other set of objections and replies.[1] Throughout his *Replies,* Descartes was scornful of Bourdin. He said of him: "[W]hen, some eighteen months ago, I saw a preliminary attack of his against me which, in my judgment, did not attempt to discover the truth but foisted on me views which I had never written or thought, I did not hide the fact that I would in future regard anything which he as an individual produced as unworthy of a reply."[2] But, in fact, Descartes did reply to Bourdin. He did so in print, having published Bourdin's objections, even though Bourdin apparently offered to treat the matter privately: if Descartes agreed not to write against the Jesuits, Bourdin would send him his objections without divulging them to anyone else.[3]

The reason Descartes gave for choosing to answer Bourdin is that it was possible that Bourdin's objections represented the views of the Jesuits as a whole, instead of merely those of an individual. As Descartes said, "But since he is a member of a society which is very famous for its learning and piety, and whose members are all in such close union with each

The research for this essay was assisted by grants from the National Endowment for the Humanities, an independent Federal Agency, and the National Science Foundation (NSF grant no. DIR-9011998).
   1. Gassendi's Fifth Objections with Descartes' replies ultimately gained the distinction of being the longest exchange, given Gassendi's first attempts at objections, Descartes' replies, Gassendi's subsequent objections—the *Disquisitio metaphysica*—and Descartes' subsequent replies included as an appendix to the 1647 French edition of the *Meditations, Objections, and Replies.*
   2. AT 7:452; CSM 2:303.
   3. AT 3:464–68.

other that it is rare that anything is done by one of them which is not approved by all, I confess that I did not only 'beg' but also 'insistently demand' that some members of the society should examine what I had written and be kind enough to point out to me anything which departed from the truth."[4]

For Descartes, if the objections were Bourdin's own views, they would not be worthy of a reply; on the other hand, if they represented the views of the Jesuits, they should be taken very seriously. How could he tell? Apparently, the only criterion Descartes had for determining whether Bourdin's objections were Bourdin's handiwork or that of the society as a whole was whether they contained solid reasons or mere rhetoric: "Now I would take the same view of the present essay, and believe that it was written at the instigation of the Society as a whole, if only I were certain that it contained no quibbles or sophisms or abuse or empty verbiage."[5] Descartes ultimately reserved judgment about whether the nature of Bourdin's response indicated that it was composed solely by Bourdin. Descartes did think that Bourdin's objections contained quibbles, sophisms, and empty verbiage, but he was not certain. He hesitated to draw that conclusion in case the objections were actually produced by the society as a whole. Not trusting his own judgment in this matter, he left it to his reader to determine whether the work was Bourdin's alone or the whole society's.

Twentieth-century readers might be able to determine this matter, since they have at their disposal some means that Descartes did not have. In the same fashion as Descartes, they would be able to look at the internal consistency and cogency of Bourdin's objections, but they would also be in a better position to tell whether Bourdin actually meant what he said; that is, they would be able to compare Bourdin's objections with his other writings for consistency. Moreover, they would be able to place Bourdin's objections in the context of seventeenth-century Jesuit intellectual life, especially given the Jesuits' behavior toward Cartesianism after Descartes' death. That is what I propose to do. I will argue that Bourdin's objections are best seen as a collective response by the Jesuits.

Bourdin versus Descartes, Round 1

Pierre Bourdin was born the year before Descartes, in 1595. He became professor of humanities at La Flèche just after Descartes left (1618–23);

4. AT 7:452; CSM 2:303.
5. AT7:453; CSM 2:303.

he returned as professor of rhetoric in 1633 and taught mathematics there in 1634. He was sent to Paris, to the Collège de Clermont, later known as the Collège Louis-le-Grand, in 1635. He stayed at Clermont until his death in 1653. On June 30 and July 1, 1640, Bourdin held a public disputation at Clermont in which his student, a young noble named Charles Potier, defended some theses; among these were three articles concerning Descartes' theory of subtle matter, reflection, and refraction. Bourdin composed a preface to the theses, a *velitatio* (or skirmish), which he delivered himself. Mersenne attended the disputation and defended Descartes. He apparently chastised Bourdin for having attacked Descartes publicly without having sent Descartes his objections; Mersenne then forwarded the *velitatio* to Descartes, together with the three articles concerning Descartes' doctrines, as if they came from Bourdin himself.[6] On a couple of occasions, Descartes asked Mersenne to tell him whether the *velitatio* sent by Mersenne was given to him by Bourdin, so that Descartes might judge whether Bourdin had acted in good faith.[7]

Descartes wrote to Mersenne on July 22, 1640, thanking him for the affection Mersenne showed for him "in the dispute against the theses of the Jesuits." He told Mersenne that he had written to the rector of Clermont College requesting that objections against his theories be addressed to him, for he "do[es] not want to have any dealings with any of them in particular, except insofar as it would be attested to by the order as a whole." And he complained that the *velitatio* Mersenne sent him was "written with the intent to obscure rather than to illuminate the truth."[8] In another letter, Descartes told Mersenne that he was shocked by the *velitatio* of Father Bourdin, for Bourdin did not have a single objection to anything Descartes had written, but rather attacked doctrines Descartes did not hold.[9] Descartes characterized the affair as his going to war with the Jesuits, adding that "their mathematician of Paris has publicly refuted my *Dioptrics* in his theses—about which I have written to his Superior, in order to engage the whole order in this dispute."[10]

It is important to note that Descartes thought of Bourdin's objections as those of the Jesuits *as a group,* in keeping with his general opinion that the Jesuits normally acted as a corporate body, that the opinion of one was likely to reflect the opinion of all. Descartes did not regard Bourdin's

6. Baillet 1691, 2:73.

7. See, e.g., AT 3:162. Descartes gave a summary of the events in the Letter to Dinet, in AT 7:566–72.

8. AT 3:94.

9. AT 3:127–28.

10. AT 3:103.

initial attack as a solitary gesture; rather, he acted as if he had just received the answer he had been waiting for, concerning whether the Jesuits would support him or not.[11] He even believed that the appearance of Bourdin's attack as that of a solitary agent acting on his own was itself a matter of conspiracy: "[H]aving recognized, in P. Bourdin's action as well as in the actions of several others, that there are many who speak of me disparagingly, and that, having no means to harm me by the power of their reasons, they have undertaken to do so by the multitude of their voices, I do not wish to address myself to any of them in particular, which would be an infinite and impossible task."[12] Descartes seemed to think that if real fault had been found with his doctrines, the Jesuits would have given their reasons officially, instead of having many, including Bourdin, appear to act as individuals.

The Bourdin affair degenerated further, with Descartes referring to Bourdin's objections as *cavillations,* that is, "quibbles" or "cavils."[13] The war imagery continued: "[T]he cavils of Father Bourdin have resolved me to arm myself from now on, as much as I can, with the authority of others, since the truth is so little appreciated alone."[14] The period of this dispute was a particularly difficult one for Descartes, since it was near the time of his publication of the *Meditations,* his work on metaphysics, which he had only sketched in the *Discourse.* The new work was certain to lead him into greater controversies. The summer of 1640 was the time when Mersenne was sending out Descartes' *Meditations,* requesting objections that would be published with the work itself. Descartes expected a set of objections from Bourdin. Bourdin wrote the Seventh Objections, which were not received by Descartes in time for the first printing of the *Meditations, Objections, and Replies* but made the second printing. The exchange was not successful. Descartes complained about Bourdin in a letter to Father Dinet, head of all French Jesuits, and generally treated his objections as silly or misguided. However, Bourdin's criticisms, though verbose, cannot be accurately described as silly. Setting aside his exchanges with Descartes, Bourdin does not generally strike one as misguided; his views cannot even be described as very conservative.

By 1640, when Bourdin debated with Descartes, he had already published three books: *Prima geometriae elementa,* following Euclid; *Geometria, nova methodo;* and *Le cours de mathématique.* He would shortly

11. Baillet (1691, 2:73–74) said that Descartes was mistaken in this. Descartes repeated the passage from his letter to Mersenne in his Seventh Replies, in AT 7:452–53; CSM 2:303.
12. AT 3:161.
13. See, for example, AT 3:163, 184, 250.
14. AT 3:184.

be publishing his fourth, *L'introduction à la mathématique*.[15] Bourdin's mathematics, like that of most Jesuits, is characterized by its practical bent. This is made clear by Bourdin's *Cours de mathématique*, which contains materials on fortification, terrain, and military architecture and sections on cosmography and the use of a terrestrial globe. It is also supported by the subject of his two posthumous publications: *L'architecture militaire, ou L'art de fortifier les places regulières et irregulières* and *Le dessein, ou La perspective militaire*.[16]

In his *Cours de mathématique* Bourdin did not shy away from discussing some dangerous philosophical topics, such as the Copernican system, whose main premise had been declared heretical by the Catholic Church in 1616 and used as the reason for the condemnation of Galileo in 1633. Bourdin simply treated the Copernican system as hypothesis, along with the Aristotelian and Tychonic systems, and took an instrumentalist line on the status of such hypotheses, as was common in mathematical works: "Since it can happen that the earth, the sun, and such parts can be disposed in various fashions and still all these appearances remain, and can be well explained, astronomers use various means of ordering and disposing the world, each constructing his own hypothesis, according to whether he judges it to be easiest, or following some new remarks he makes, seeking nothing other than its usefulness in explaining the appearances of the world."[17] The instrumentalist preamble to the *Cours de mathématique* was followed by a section on the hypothesis of the ancients, that is, homocentric spheres, plus epicycles and solid eccentrics; then by a section on the Copernican hypothesis, with a reference to sunspots as stars revolving around the sun; and finally by a section on the Tychonic hypotheses, including a reference to Galileo's discovery of the moons of Jupiter.[18] In his section on optics, Bourdin even gave a positive reference to the telescope—*les lunettes extraordinaires*—as an instru-

15. Bourdin 1639 (divided into *geometria speculativa, geometria practica, notae geometricae,* and *aditus in arithmetica*); 1640; 1661. "*L'introduction à la mathématique* [1643] contenant les coignaissances, et pratiques necéssaires à ceux qui commencent d'apprendre les mathématiques. Le tout tiré des élémens d'Euclide rengez et demonstrez d'une façon plus briève, et plus facile que l'ordinaire. Part I: géometrie; II: géometrie de raison; III: abrégé de l'arithmétique." The reason I include *Le cours de mathématique* among those works published by Bourdin on or before 1640 is that an anonymous 1645 version was identified as a revised edition; the 1645 work contains plates dated 1631. Thus 1631 is probably the date of the first edition, with printings in 1640 and 1641. See P. J. Jones 1947, 119–20.

16. Bourdin 1655a, 1655b.

17. Bourdin 1661, 124.

18. Ibid., 126, 128, 130. Sunspots had been widely interpreted, in France, as small planets going around the sun; see Baumgartner 1987.

ment for the discovery of astronomical effects.[19] It should be added, however, that Bourdin's view of optics was rudimentary: he discussed reflection, saying that the angle of incidence equals the angle of reflection, and refraction, saying that refraction bends light toward the normal when the light passes from a rarer to a denser medium, but he did not give the Snell-Descartes sine law.[20]

Despite his instrumentalism, Bourdin seems to have had a preference for the Tychonic system, calling it "the one in fashion today, having been sketched by Martianus Capella and polished and completed not long ago by Tycho Brahe, that excellent mathematician."[21] There is another reason for thinking that Bourdin followed the fashion for the Tychonic system. In his one public departure from the realm of mathematics as defined in the seventeenth century, into the realms of physics and cosmology, Bourdin gave arguments and sketched doctrines compatible only with the Tychonic system. Bourdin's cosmological work consists of a single volume in which two small treatises on the same general subject are bound together: *Sol flamma* and *Aphorismi analogici*.[22] In these works, Bourdin argued that the sun is a blazing fire, a position inconsistent with the Aristotelian theory of the heavens, as Bourdin knew quite well,[23] and supported by such innovators as Descartes. He even referred to Descartes as someone who held the position: "novissime a Renatus des Cartes solem docet esse flammam."[24] Bourdin's basic argument is that the sun is a body on which there are sunspots and small torches, as the telescope rendered evident. Thus the sun is corruptible matter, not incorruptible ether, as Aristotle had it.[25] In the *Aphorismi analogici,* such considerations compelled Bourdin to adopt a Tychonic cosmology. He moved from an explanation of sunspots by analogy with foam bubbling up from the sea, to

19. Bourdin 1661, 176.
20. Ibid., 156–86.
21. Ibid., 130.
22. Bourdin 1646.
23. Bourdin 1646, *Sol flamma* 1–3: "1. auctores, et argumenta sententia negantis [Aristoteles]."
24. Bourdin 1646, *Sol flamma* 5. This seems to be the only reference to Descartes in Bourdin's published works, other than in the Seventh Objections. The author to whom Bourdin was most indebted was Christopher Scheiner, the Jesuit astronomer, whose 1630s *Rosa Ursina* is referred to extensively in *Sol flamma.*
25. Bourdin 1646, *Sol flamma:* "sol est corpus; in quo sunt eiusmodi maculae, et faculae, ut patet ex telescopio, et parallaxi, quae docet haec omnia non distare a sole; ergo sol est corruptibilis" (7–9); "atqui sol paret flamma (ut patet rescipiendi per telescopium; quo, ut docet Scheiner lib. 2. *Rosa Ursina,* cap. 4. deprehenduntur in sole multa flammae signa)" (14–16).

there being three regions of stars and planets, to magnetic phenomena affecting both the earth and the heavens.[26] But he rejected the Copernican hypothesis, claiming that the earth stays still.[27]

From Bourdin's other writings, it would be difficult to infer that Bourdin would be a dogmatic opponent of Descartes. Regardless, Descartes treated Bourdin as an unworthy critic, insulting the objector and evading his objections: "[W]hat he does is to take fragments from my Meditations and ineptly piece them together so as to make a mask which will not so much cover as distort my features."[28] Descartes compared his reasoning with that of children: "I am amazed that his ingenuity has been unable to devise anything more plausible or subtle. I am also amazed that he has the leisure to produce such a verbose refutation of an opinion which is so absurd that it would not even strike a seven year old child as plausible."[29] He sneered at Bourdin: "[H]e is foisting on me, good-natured fellow that he is, a piece of reasoning that is worthy of himself alone," and "he finally reaches a conclusion which is wholly true when he says that in all these matters he has 'merely displayed his weakness of mind.'"[30] Descartes also overlaid his insults with the suggestion that Bourdin was not actually inept, but just pretending to be so—that he was playing the clown: "Yet it is embarrassing to see a Reverend Father so obsessed with the desire to quibble that he is driven to play the buffoon. In presenting himself as hesitant, slow and of meagre intellect, he seems eager to imitate not so

---

26. Bourdin 1646, *Aphorismi analogici:* "Explicantur maculae solis exemplo spumarum maris" (44–46); "Distinguuntur stellae et planetae in tres partes seu regiones" (49–50); "De influxu magnetico mundi tum caelesti tum terrestri' (50–52); "De terminis fluxus magnetici mundi" (52–53).

27. Ibid., 65–66: "Terra quies probatur primo." The Jesuits seem to have widely accepted the cosmology of three heavens by the 1640s. In a 1642 thesis by Jean Tournemine, a student at La Flèche, we are told that "apostolic authority teaches us that there are three heavens. The first is that of the planets, whose substance is fluid, as shown by astronomical observations; the second is the firmament, a solid body as the name indicates; and the third is the empyrean, where all species of stars are to be distinguished" (in Rochemonteix 1889, 4:365–68). One can find a similar theory in Guillelmus de la Vigne's 1666 thesis from the Jesuit college at Caen, the Collège du Mont (Caen MS 468). The reason that this theory of the heavens seems to be Tychonic is that solidity is attributed to the firmament, or the outermost heavenly body, containing the fluid universe of the planets.

28. AT 7:454; CSM 2: 304. For an account of Descartes' evasions, see Ariew 1994.

29. AT 7:466; CSM 2:313. Later on, the comparison is with a three year old: "A three year old child could supply the answer to this" (AT 7:514; CSM 2:350).

30. AT 7:474; CSM 2:319. AT 7:477; CSM 2:321. Also: "These comments are amusing enough, if only because they would be so inappropriate if they were intended to be serious" (AT 7:511; CSM 2:348); "Having asked this utterly absurd question" (AT 7:524; CSM 2:356).

much the clowns of Roman comedy like Epidicus and Parmenon as the cheap comedian of the modern stage who aims to attempt to raise a laugh by his own ineptitude."[31] In the end, Descartes accused Bourdin directly of being a liar:

> The conclusion, unless I am wholly ignorant of what is meant by the verb "to lie," is that he is inexcusably lying—saying what he does not believe and knows to be false. Although I am very reluctant to use such a distasteful term, the defence of the truth which I have undertaken requires of me that I should not refuse to call something by the proper word, when my critic is so unashamedly and openly guilty of the deed. Throughout this whole discussion he does virtually nothing else but repeat this foolish lie in a hundred different ways, and try to persuade and bludgeon the reader into accepting it.[32]

Bourdin was treated roughly by Descartes, but Descartes treated many of his other objectors in a similar fashion, though he was not as insulting to the others. Perhaps Bourdin merited the treatment. He was not very efficient in his criticism. Part of the problem may be Bourdin's attempt at rhetoric, that is, his decision to write his objections in dialogue form and his frequent rhetorical flourishes. That decision was disastrous, given that Descartes had the last word and undercut Bourdin's objections by interspersing his own replies within Bourdin's dialogue form, making the Seventh Objections and Replies extremely difficult to read. Bourdin's objections also suffer because Descartes, a master at rhetorical technique, did not fail to use all his skill in his even longer replies. All this should make anyone extremely uneasy when attempting an analysis of the Seventh Objections and Replies. In fact, Descartes himself in his dealings with Bourdin was sometimes unsure that he had understood the thrust of his interlocutor's objections. In a revealing passage from a letter to Mersenne, Descartes said: "I wish to believe that Father Bourdin did not understand my demonstration," but that does not prevent his objections from "con-

---

31. AT 7:492–93; CSM 2:333. Also AT 7:510; CSM 2:347. Also: "[A]s my critic here jeeringly and impertinently suggests" (AT 7:491; CSM 2:332); "My critic's next piece of reasoning is also well suited to the role he has assumed. . . . Remarks that are manifestly absurd are well suited to provoke laughter. . . . My critic continues to play his comic role outstandingly well when he tells the story of the peasant. But what is most laughable here is that he thinks the story applies to my words, when in fact it applies only to his own" (AT 7:510; CSM 2:347); "There is much here that deserves to be laughed at now and for evermore, but rather than point this out I prefer to respect the actor's costume that my critic has assumed; and indeed I do not think it is right for me to spend all this time laughing at such ill-considered comments" (AT 7:517; CSM 2:350).

32. AT 7:525; CSM 2:357.

taining cavils that were not merely invented through ignorance, but because of some subtlety that I do not understand."[33] Still, there are some points to be made about the structure of Bourdin's attack on Descartes.

Bourdin versus Descartes, Round 2

Bourdin's objections were all directed against Descartes' method of doubt. He clearly hoped to derail Descartes' whole enterprise from the start. His strategy was to show that the method failed either because it was untrue to itself and smuggled in various external principles or because, if the method did not smuggle anything in, it went nowhere. The second half of the disjunction was further subdivided, and both halves of Bourdin's dilemma manifested themselves in different ways. Here is a clear statement of the first horn: "[The method] sins by omission. For having laid it down as a fundamental principle that we should 'take great care not to admit as true anything we cannot prove to be true,' it proceeds to violate the principle more than once. It assumes with impunity, and regards as completely certain and true, unproved statements such as 'The senses sometimes deceive us,' 'We are all dreaming,' 'Some people go mad,' and so on."[34] Bourdin's claim is that the principle of doubt is itself a principle; but since it is not clear that there is anything wrong with smuggling in various principles, Bourdin also had to argue that the principles Descartes smuggled in were defective in some way. First he argued that Descartes' principles are not as certain as the common principles denied by the method of doubt: "[L]et me come to your maxim 'If something appears certain to someone who is in doubt whether he is dreaming or awake, then it is certain—indeed so certain that it can be laid down as a basic principle of a scientific and metaphysical system of the highest certainty and exactness.' You have not at any point managed to make me consider this maxim to be as certain as the proposition that two and three make five."[35]

---

33. AT 3:249–50.

34. AT 7:532; CSM 2:363. Here is another version of the first horn: "You acknowledge that all your former beliefs are doubtful, and you claim that you are compelled to admit this. . . . , What compels you, may I ask? I heard you say just now that you had 'powerful and well thought-out reasons.' But what are they? And if they are powerful, why renounce them? Why not keep them? If, on the other hand, they are doubtful and completely suspect, how have they managed to force or compel you?" (AT 7:469; CSM 2:315). See also AT 7:504; CSM 2:342.

35. AT 7:457; CSM 2:306–7. Also: "[S]ince you are opening a new school of philosophy and thinking of your disciples, is it that you want this inscription to be placed over the door in gold letters: 'I cannot go too far in my distrustful attitude'? Will the students who enter

It should be noted that Bourdin did not take the skeptic's gambit against the skeptic. That is, when Bourdin tried to show that residual principles had been smuggled in, he did not object to those principles by using the principle of doubt reflexively against them; rather, he simply pointed out that the principles smuggled in were not as worthy or as certain as the common principles ruled out by the method. Here again is the same argument, but this time it is followed by the additional complaint of vicious circularity, that is, that what was smuggled in is the very conclusion at stake:

> But why did you not warn me at the outset? When you saw that I was all keyed up and ready to renounce my former beliefs, why did you not tell me to keep this one belief, to take from you a special certificate bearing the words "Thinking is something that belongs to the mind"? Well, I must take all the credit for warning you to stress this maxim when dealing with your next batch of beginners: take care to remind them not to renounce it along with such former beliefs as "Two and three make five." . . . You promise us that you will establish by strong arguments that the human soul is not corporeal but wholly spiritual; yet if you have presupposed as the basic premiss of your proofs the maxim "Thinking is a property of the mind, or of a wholly spiritual and incorporeal thing," will it not seem that you have presupposed, in slightly different words, the very result that was originally in question?[36]

Bourdin even supported his complaint by showing that the case was not merely hypothetical, but that there were philosophers who thought that

---

your precincts be told to lay aside the old belief that 'Two and three make five,' but to retain the maxim 'I cannot go too far in my distrustful attitude'? . . . Do you really have no scruples about renouncing such long-standing beliefs as the following: 'I have a clear and distinct idea of God'; 'Everything that I clearly and distinctly perceive is true'; or 'Thinking, nutrition and sensation do not in any way belong to the body but belong to the mind'?" (AT 7:471; CSM 2:317).

36. AT 7:489–90; CSM 2:331. Similarly: "If you are dreaming that thinking extends more widely, does it follow that it really does so? If you wish, I am ready to dream that dreaming extends more widely than thinking. But how do you know that thinking extends more widely, if there is no thinking, but only dreaming? Suppose that whenever you have thought that you were thinking while awake, you were not really thinking while awake, but merely dreaming that you were thinking while awake? What if dreaming is a single operation which enables you sometimes to dream that you are dreaming, and at other times to dream that you are thinking while awake? . . . How do you know you are either a body or a mind, since you have renounced both body and mind? What if you are neither a body nor a mind but a soul, or something else? How am I to know the answer?" (AT 7:494–95; CSM 2:334–35). See also AT 7:496; CSM 2:336.

thinking is a property of the body, so that their position cannot have been ruled out without a substantive principle: "In case you think I am throwing in this objection just for the sake of it, there are many people, and serious philosophers at that, who claim that the brutes think, and hence regard thought as an attribute which, though not common to all bodies, is common to all extended souls of the kind which the brutes possess; and this implies that it is certainly not a property that uniquely and necessarily belongs to a mind or spiritual substance!"[37]

With the second horn of the dilemma, Bourdin tried to show that the method produced nothing or that it proved too much (with the added problem that the excess produced was unreliable). Here are both sub-cases:

> The method is faulty in its principles, which are either non-existent or unlimited. Other systems which aim to derive certain results from certain starting points lay down clear, evident and innate principles. . . . But your method is quite different, since it aims to derive something not from something but from nothing. It chops off, renounces and forswears all former beliefs without exception; . . . it struggles to derive something from nothing, only to end up producing nothing at all.[38]

He followed this by arguing that the method ultimately could not have produced anything since it cut itself off from traditional means of argumentation and rejected any major premise whatever:

> The method is faulty in the implements it uses, for as long as it destroys the old without providing any replacements, it has no implements at all. Other systems have formal logic, syllogisms and reliable patterns of argument, which they use like Ariadne's thread to guide them out of the labyrinth; with these instruments they can safely and easily unravel the most complicated problems. But your new method denigrates the traditional forms of argument, and instead grows pale with a new terror—the imaginary fear of the demon which it has conjured up. It fears it may be dreaming; it has doubts about whether it is mad. If you propose any syllogism, it will be scared of the major premiss, whatever it may be.[39]

---

37. AT 7:490; CSM 2:331.
38. AT 7:527–28; CSM 2:358–59.
39. Bourdin continues: "What will you do when your method obstinately maintains that any conclusion you draw is doubtful unless you previously know for certain that you are not dreaming or mad, and that there is a God, a truthful God, who has the evil demon

Bourdin also argued that what the method produced was unreliable:

> The method goes astray by failing to reach its goal, for it does
> not attain any certainty. Indeed, it cannot do so, since it has itself
> blocked off all the roads to the truth. . . . Here then are the chief
> ways in which your method cuts its own throat or cuts off all
> hope of attaining the light of truth. (1) You do not know whether
> you are dreaming or awake, and hence you can place no more
> confidence in your thoughts and reasonings (that is, if you have
> any, and are not merely dreaming you have them) than a dreamer
> can place in the thoughts he has while asleep. Hence everything
> is doubtful and shaky, and your very inferences are uncertain. . . .
> (2) Until I know that there exists a God who will curb the evil
> demon, I must continue to doubt everything and consider every
> proposition as suspect. . . . So everything is doubtful, just as we
> found under point (1), and hence we have nothing left which will
> be the slightest use for investigating the truth.[40]

Finally, he argued generally that the method was quixotic and imprudent:

> The method goes astray by being excessive. That is, it attempts
> more than the laws of prudence demand of it, more, indeed, than
> any mortal demands. Admittedly there are people who are look-
> ing for a demonstration of the existence of God and the immor-
> tality of the human mind. But you will not find anyone up till
> now who has been dissatisfied if propositions like "God exists
> and the world is governed by him," or "The souls of men are
> spiritual and immortal," are known with as much certainty as
> "Two and three make five," or "I have a head and a body." So all
> these efforts to search for some higher grade of certainty are
> superfluous.[41]

Whatever Descartes may have thoght about Bourdin's criticism, at least
Bourdin's attack is consistent with Jesuit pedagogical practice.[42] By restricting himself to a critique of Descartes' method, Bourdin did not have

---

under control? You may produce your syllogism 'To say that something is contained in the
nature or concept of anything is the same as saying that it is really true of that thing; now
existence is contained in the nature and concept of God,' etc. But what if your method
repudiates both the content and the form of this argument and other arguments of this
kind? Whatever argument you press, the reply will be 'Wait until I know there is a God and
see the demon curbed.' . . . In short, I have just one point to make: if you take away all
form, nothing remains but the formless, or the deformed" (AT 7:528–29; CSM 2:359–60).

40. AT 7:529–30; CSM 2:360–61.

41. AT 7:530; CSM 2:361.

42. See Ariew 1992b, and forthcoming.

to engage any particular doctrinal point; instead he emphasized the difficulty that Jesuits would have with any method that espoused skepticism, even if only as a preliminary step. There seems to have been broad agreement between Bourdin's concerns and subsequent Jesuit (or other) condemnations of Cartesianism, though the matter might have been obscured by the doctrinal nature of official condemnations.

Bourdin versus Descartes, Round 3

Most of the condemnations of Cartesianism from 1642 to 1691 simply rejected some aspects of it that seemed incompatible with standard explanations of theological matters—the mystery of transubstantiation in the sacrament of the Eucharist, for example. But here and there one can find other kinds of problems with Cartesianism. I will first sketch some condemnations of Descartes dealing with doctrinal problems and then look at some that concentrate on pedagogical and pragmatic difficulties.

The condemnation of Cartesianism at Louvain in 1662 (which, according to Victor Cousin, was instigated by Jesuits) refers only to five difficulties with Cartesian doctrine: the definition of substance (*Principles of Philosophy* 1.51–52), the rejection of substantial forms or real accidents existing by themselves (Sixth Replies), extension as an essential attribute of substance (*Principles* 1.53), the indefinite extension of the world (*Principles* 2.21), and the plurality of worlds (*Principles* 2.22).[43] The first three difficulties were repeated in slightly different form, with some rhetorical elaborations, in a disputation by the Jesuits of Clermont College during 1665:

> To say no more, the Cartesian hypothesis must be distasteful to mathematics, philosophy, and theology. To philosophy because it overthrows all its principles and ideas which commonsense has accepted for centuries; to mathematics, because it is applied to the explanation of natural things, which are of another kind, not without great disturbance of order; to theology, because it seems to follow from the hypothesis that (i) too much is attributed to the fortuitous concourse of corpuscles, which favors the atheist; (ii) there is no necessity to allow a substantial form in man, which favors the impious and dissolute; (iii) there can be no conversion of bread and wine in the Eucharist into the blood and body of Christ, nor can it be determined what is destroyed in that conversion, which favors heretics.[44]

43. Argentré 1736, pt. 2, pp. 303–4.
44. Oldenburg 1966, 2:435. I am indebted to Mordechai Feingold for this reference.

The same general set of Cartesian doctrines was prohibited in 1678 during an assembly between the Jesuits and the Oratorians:

> In physics one must not stray from Aristotle's physics nor the principles of Aristotle's physics commonly received in the colleges in order to follow the doctrine of M. Descartes, which the king has forbidden for good reasons.
>
> One must teach 1. That actual and external extension is not the essence of matter. 2. That in each natural body there is a substantial form really distinct from matter. 3. That there are real and absolute accidents inherent in their subjects, which can supernaturally be without any subjects. 4. That the soul is really present and united to the whole of the body and to all parts of the body. 5. That thought and knowledge are not the essence of the rational soul. 6. That there is no repugnance in God's creating several worlds at the same time. 7. That the void is not impossible.[45]

The Cartesian doctrines rejected at Louvain and by the Jesuits and Oratorians also came up in other condemnations during the same time. The king's decree of 1671 against Cartesianism, transmitted by the archbishop of Paris, is concerned with the question of the Eucharist and thus alludes to the condemnation of 1624 against the atomists' denial of real qualities and substantial forms.[46] The difficulties that Oratorians had at Angers during 1675–78 also reflect a general disapproval of the same set of Cartesian doctrines. For example, Fathers Fromentier, L'Amy, and Villecroze taught "the opinion of the Cartesians who state that there are no species or real accidents in the Eucharist, which is contrary to the theology of the Holy Fathers and to the doctrine of the church . . . and which was censured by the Sorbonne in 1624 as bold, erroneous, and approaching heresy."[47] Similarly, the Fathers taught that "the world is infinite in its extension, a principle that is no less dangerous than the first."[48]

The censure of the Oratorian professors also contains other nondoctrinal or pedagogical and pragmatic elements. The Oratorians were reproached for allowing their students to have "published a thesis without having submitted it to examination [by the rector]."[49] And they were re-

45. *Concordat entre les Jesuites et les Pères de l'Oratoire*, Actes de la Sixième Assemblée, Sept. 1678, in Bayle 1684, 11–12.

46. See Ariew 1992a.

47. Babin 1679, 39, 44.

48. Ibid., 40. The sentence continues, "it is true that the Cartesians do not make use of the word *infinite*, which would be too odious, but only of the word *indefinite*, which is the same thing and adds only a single syllable to what we are saying about the infinite."

49. Ibid., 24.

proached for not having rejected Descartes' skeptical method: "To say that one must doubt all things is a principle that tends toward atheism . . . or at least toward the heresy of the Manicheans." [50] Later on, criticisms of Descartes' method of doubt à la Bourdin became standard in Scholastic textbooks. Here is a typical example:

> But it is clear that if for once one doubts seriously and effectively everything, it becomes impossible to be certain of anything, no matter what examination one conducts, because, if one can become certain of something, after such a serious doubt, it would only be through the evidence of the thing, since there is no other rule of human certainty than the evidence of the thing, according to the Cartesians. Now, we assume that they do doubt seriously the most evident things, even their own thought and their own existence, and that consequently, after such a general and serious doubt, it would be impossible to become certain of anything, whatever examination is conducted of it. That is why the Cartesians are to be distinguished from the Pyrrhonists, in that they do not reason well when they say that after such a general doubt one can become certain of something, while the Pyrrhonists reason well and in conformity with their principles when they say that one cannot be certain of anything after doubting everything. [51]

The principal reason for asserting that Bourdin's objections are consistent with Jesuit attitudes emerges more clearly if one begins to focus on nondoctrinal elements in condemnations of Cartesianism by the Jesuits. This is not to say that the Jesuits alone were unhappy about the pedagogical and pragmatic consequences of Descartes' philosophy. In fact, as early

50. Ibid., 41. There is also an interesting attack on Descartes' *cogito* as a principle of knowledge, reminiscent of Bourdin's strategy against Descartes. The author of the objection can no more accept the *cogito* as a principle in the framework of Aristotelian epistemology than Bourdin could accept Cartesian doubt within that framework: "But he does not perceive that it is impossible for the argument to be the first principle of reasoning and knowledge, otherwise a thing would be a principle of itself. Now, *cogito ergo sum* is an argument, but a truly defective one, since the consequence of this enthymeme is the same thing as the antecedent. For *cogito* means in philosophical terms *ego sum cogitans*. . . . The first principle of the sciences must be universal and necessary, because science is *universaliter et necessaria;* and this principle *cogito* is something singular and extremely uncertain, since it is . . . contingent" (42).

51. Duhamel 1692, chap. 4: "Si après un doute general on peut s'assurer de quelque chose" (p. 8). See also Vincent 1677; and Huet 1689. The Jesuit view is also nicely captured by the following comment: "A la vérité [Descartes] enseigne trops à douter: et ce n'est pas un bon modele à des esprits naturellement crédules: mais enfin il est plus original que les autres" (Rapin 1725, 366).

as the dispute between Scholatics and Cartesians in the condemnation of
Cartesianism by the academic senate of Utrecht, in March 1642, a
mixture of pragmatic, pedagogical, and doctrinal elements was evident.
The reasons for condemning Cartesianism move from pragmatic to
pedagogical to doctrinal. Here is the Utrecht edict, as quoted by Descartes himself:

> The professors reject this new philosophy for three reasons. First,
> it is opposed to the traditional philosophy which universities
> throughout the world have hitherto taught on the best advice,
> and it undermines its foundations. Second, it turns away the
> young from this sound and traditional philosophy and prevents
> them [from] reaching the heights of erudition; for once they have
> begun to rely on the new philosophy and its supposed solutions,
> they are unable to understand the technical terms which are com-
> monly used in the books of traditional authors and in the lectures
> and debates of their professors. And, lastly, various false and ab-
> surd opinions either follow from the new philosophy or can be
> rashly deduced by the young—opinions which are in conflict
> with other disciplines and faculties and above all with orthodox
> theology.[52]

By 1691, when the University of Paris finally condemned Cartesianism,
the focus was no longer only on Cartesian doctrines as such. Much of the
Paris edict is given over the condemnation of the Cartesian method of
doubt, with the following propositions prohibited:

> 1. One must rid oneself of all kinds of prejudices and doubt ev-
> erything before being certain of any knowledge. 2. One must
> doubt whether there is a God until one has a clear and distinct
> knowledge of it. 3. We do not know whether God did not create
> us such that we are always deceived in the very things that appear
> the clearest. 4. As a philosopher, one must not develop fully the
> unfortunate consequences that an opinion might have for faith,
> even when the opinion appears incompatible with faith; notwith-
> standing this, one must stop at that opinion, if it is evident.[53]

52. AT 7:592; CSM 2:393.
53. Argentré 1736, pt. 1, p. 149. The rest of the Paris condemnation contained the prohi-
bition of the important Cartesian proposition "The matter of bodies is nothing other than
their extension, and one cannot be without the other." It also prohibited several propositions
not normally associated with Cartesianism: "6. One must reject all the reasons the theologians and the philosophers have used until now (with Saint Thomas) to demonstrate the
existence of God. 7. Faith, hope, and charity and generally all the supernatural habits are
nothing spiritual distinct from the soul, as the natural habits are nothing spiritual distinct

Similarly, when in 1706 the general of the Jesuit order condemned thirty different Cartesian propositions, he did not fail to include some against Descartes' method of doubt. Prohibited were the following propositions:

> 1. The human mind can and must doubt everything except that it thinks and consequently that it exists. 2. Of the remainder, one can have certain and reasoned knowledge only after having known clearly and distinctly that God exists, that he is supremely good, infallible, and incapable of inducing our minds into error. 3. Before having knowledge of the existence of God, each person could and should always remain in doubt about whether the nature with which one has been created is not such that it is mistaken about the judgments that appear most certain and evident to it. 4. Our minds, to the extent that they are finite, cannot know anything certain about the infinite; consequently, we should never make it the object of our discussions. 5. Beyond divine faith, no one can be certain that bodies exist— not even one's own body.[54]

There is no doubt that during the seventeenth century the Jesuits became the enemies of Cartesian philosophy and science; the condemnations of Cartesianism, whether by the Jesuits directly or indirectly, attest to that fact. There is also no doubt that the Cartesian philosophy was rejected by the Jesuits (and other Catholics) on doctrinal grounds. However, the greatest problem the Jesuits (though not the Jesuits exclusively) had with Cartesian philosophy and science does not represent a doctrinal conflict, but a conflict over pedagogy and other pragmatic matters. Even if Descartes' doctrines did not clash with Jesuit doctrines, Descartes' philosophy simply could not have failed to clash with other Jesuit intellectual attitudes. Descartes was offering a novel philosophy, and even worse, his novel philosophy was based on a method that espouses initial doubt. The Jesuits could not accept a method of doubt or skepticism even as a heuristic. This is amply demonstrated in the dispute between Descartes and Bourdin and the Jesuits' condemnation of Cartesianism. Bourdin himself was not averse to all novel philosophies, yet he rejected

---

from mind and will. 8. All the actions of the infidels are sins. 9. The state of pure nature is impossible. 10. The invincible ignorance of natural right does not excuse sin. 11. One is free, provided that one acts with judgment and with full knowledge, even when one acts necessarily."

54. "Prohibited Propositions by Michel-Angelo Tamburini, General of the Order in 1706," in Rochemonteix 1889, 4:89–90n. For an English translation of the other prohibited propositions, see Ariew 1992b, 89–90.

Cartesian philosophy and its method of doubt. Bourdin's rejection of Cartesian skepticism could not have been an individual quirk. The Seventh Objections must have been the reply Descartes was waiting for, about whether the Jesuits would support his philosophy. The answer was negative.

EPILOGUE

Until recently, as Roger Ariew and I remarked in the Prologue to this volume, the history of philosophy has consisted chiefly in the examination, more or less analytic, of our major texts. Those very texts, however, it now appears to many scholars, including the contributors to this collection, can be freshly illuminated if the perspective for their study is broadened, so that we see their authors in the context not so much of our as of their contemporary concerns. In this concluding essay I hope to underscore this thesis, first, by saying a little more about its application to Descartes, with an example drawing particularly on one of our essays, and then by setting the Cartesian dialogic enterprise into a broader historical context, comparing it with the first corpus of dialogues in the history of Western thought, and suggesting that we consider philosophy itself as well as the history of philosophy from this perspective.

Cartesian Dialogue

There is a sense in which, internally, the *Meditations* constitute a dialogue: the debate between the Meditator, as he advances in his effort to free his mind from entanglement with the senses, and the old, confused "self" he is endeavoring to leave behind. This plurality of voices in the *Meditations* themselves is mentioned, for example, in the opening address of J.-M. Beyssade to the 1992 Paris colloquium on the *Replies,* which formed a sequel to our conference on the *Objections.*[1] Important though it is in any judicious reading of the *Meditations,* however, we may ignore that inner dialogue here, as yet another way of looking at the text in isolation, and start from the conclusion of Jean-Luc Marion's essay, which served, fittingly, as the keynote address for our conference of 1992. It would be "illegitimate," he says, "to read the *Meditations* in abstraction from the *Objections* and *Replies,* with which they intentionally form an

---

1. Beyssade 1994. See also Garber 1986, as well as Curley 1986 and Rodis-Lewis 1986.

organic whole." Further, he argues persuasively, "it would also be wholly illegitimate to read them otherwise than as replies to the objections evoked by the *Discourse on Method*." And he concludes: " Far from being soliloquy or solipsism, Cartesian thought, insofar as it obeys a logic of argumentation, is inscribed in its origin in the responsorial space of dialogue." Beyssade, in discussing this thesis at the *Replies* conference, places Marion's approach at the opposite extreme from a reading like that of Charles Adam, who took it that Descartes had definitively formulated his metaphysic by 1630 and needed no outside help to elaborate, let alone to alter, it. Clearly, however, to judge from his own thoughtful exposition, Beyssade himself accepts something in the direction of the Marion thesis as appropriate for the study of Cartesian texts. He points out how, to put it in Descartes' own words, he "changed, corrected, and added to" his first philosophy in the course of the reception of the *Objections* and the formulation of the *Replies*.[2]

This general thesis is reinforced in many of the preceding essays. For example, Armogathe has enriched our understanding of Descartes' (indirect) exchange with the obscure Dutch priest Caterus by delving into the opinions of Caterus' teachers. Garber argues that the Second Objections, collected by Mersenne, very probably derived in part from J.-B. Morin, whose *Quod Deus Sit* purported to provide the kind of a priori proof of God's existence that the writers of the Second Objections called for. And in fact, Garber tells us, this dialogue was continued with Morin's reply to Descartes' response to his challenge. Carraud shows how Arnauld's work before he read Descartes' manuscript contributed to the ease with which his thought was "Cartesianized," and how, also, his reading of St. Augustine himself was altered and enriched through this experience. And so on—it would be tedious to recapitulate further. Instead, I want to reflect for a moment on the theme of one of our essays: Menn's account of Descartes' denial of real qualities, or rather, on one of the questions raised by Menn's rich and scholarly account, that is, the distinction between accidents and modes.

If we look at the *Meditations* in any way at all in historical context, we are likely to see their author either (as he sometimes claimed) as our liberator from a stultified and stultifying tradition or as a radical reformer within that tradition. From the latter perspective, he appears as a major figure within the tradition of substance metaphysics, the tradition founded by Aristotle and carried forward in numerous transformations until almost our own day. A founding principle of such a metaphysic is

2. Beyssade 1994.

the thesis that everything there is is either a substance (existing independently or "in itself") or in some way a qualifier of a substance, an "accident." But if Cartesian metaphysics has its roots in an Aristotelian soil, it is by no means directly Aristotle's own metaphysics from which it springs. Descartes' teachers and their teachers for many generations had been modifying and complicating Aristotle's first philosophy, and those modifications and complications are reflected in the development of Cartesian thought. To mention one obvious example, Descartes attacks substantial forms (except for the human case), but there is no such concept in the Aristotelian texts. What emerges clearly from Menn's exposition is that Descartes is also developing his metaphysical position in the context of a modified theory of accidents, or modifiers of substance, as well as of substance itself.[3]

The modification in question is the distinction between accidents and modes, a distinction current among the Scholastics but absent from the texts of Aristotle. Indeed, on a first, or even nth reading, it seems that *mode* is just Descartes' choice of a word for *accident*. And it is true that in the *Meditations* he does seem to use those two terms indifferently.[4] By the time he publishes the *Principles of Philosophy*, however, he has definitely chosen *mode* as the term to designate both special qualifications of *res extensa* and of *res cogitans*. Thus when I think about the book before me, the book's solidity is a mode of body as my present thought about the book is a mode of mind. Yet in the reply to Arnauld quoted by Menn in the opening of his essay it is plain that Descartes understood that *accident* and *mode* had both a different reference and a different sense: "[H]e thinks that I do not admit real accidents, but only modes, which cannot be understood, and indeed cannot exist, without some substance in which they are present."[5] Descartes is here reporting an objection by Arnauld, which in turn refers to his reply to Caterus about the modal character of motion.[6] Now, Arnauld had said, on the basis of that passage in the First Replies, "the author thinks there are no sensible qualities."[7] But in the reply to Caterus Descartes was discussing motion and shape, not qualities such as color, fragrance, and the like, which philosophers were later to refer to as secondary qualities, and which, in that later tradition, would often be thought to have no reality outside the mind. Why did Arnauld read this discussion of motion and shape as a denial that there are any sensible qualities, and why did Descartes in his reply turn this statement

3. See Garber 1992, 69–70, 329n24.
4. Ibid., 327–28n6.
5. AT 7:248.

6. See AT 7:121.
7. AT 7:217.

into a denial on his part of "real accidents" as distinct from modes, which cannot exist in independence of the substance they modify? To us, inheritors of the primary-secondary distinction, it seems that a shape, for example, would have a better chance of existing "independently" (apart from an observer) than a color or taste or sound. To those who have sat at Eddington's table (or have thought they did), or who recall the distinction between the "manifest" and "scientific" image, this seems almost backward. Yet Descartes is simply applying to his own text a pair of concepts that any of his learned readers would have recognized. Not only (as Suárez admitted) because of its application to the Eucharist, but in terms of Scholastic theories of predication, there seemed to be a difference between predicates that were *res* and those that were not, or could not be. The former are real qualities, the latter modes.

What Descartes wanted to do, as he made clear in a letter to Mersenne quoted by Menn, was to generalize the concept of mode to all characteristics of bodies: there are none that are *res*.[8] And so it turns out that it would be best to use the term *mode* to describe what would once have been called the accidents of bodies. For "accidents" include real accidents, which add a *res* to the substance in which they inhere, like, in Ockham's list, white or black, or hot or cold. "Modes," however, add no "real" entity to their subject. On the other hand, in contrast to ordinary accidents, they modify the very essence of their subject. To be a sphere is to be an extended thing of just this variety. To heat it or cool it, darken it or lighten it, is not to alter it in its essential nature. So modes are in two ways closer to their "underlying" substances than other accidents. They depend wholly for their very entity on their subjects, and so are less "real" than those so-called "real qualities." But they are by the same token closer to their subjects' natures than are the other accidents, however "real": they are modifications of the subject's very nature, not extras added on to it "like little souls to their bodies."[9] Of those extras Descartes has no need; he can do very well with modes only.

Are modes, as distinct from real accidents, uniquely quantitative or at least quantifiable? Certainly modes of Cartesian bodies (or body) are so. The passage Menn cites from Ockham also suggests something of this sort, since it is when we cannot explain successive predicates through local motion that we need a *res*. But as Menn points out in a note, Ockham did not accept the concept of mode. Moreover, the example Suárez uses in his account of modes, inherence (in quantity, not directly in substance)

8. See AT 3:648–49.
9. AT 3:648.

is clearly not quantifiable—it just is or is not.[10] This is a question that merits further exploration.

What is of interest here, however, is the relation of Descartes' account of nature to Scholastic analyses of predication. The *Meditations* are not only a dialogue of Descartes with himself, in which he moves dialectically through a series of positions from persuasion to science. They are also a dialogue with his contemporaries, a dialogue that develops through the *Objections* and *Replies*.[11] At one level this is obvious: in the case I have been discussing, Descartes receives objections from Caterus and replies; Arnauld raises an objection based on that reply, to which Descartes in turn replies. But we can also see in the course of these exchanges the elaboration of the metaphysic Descartes is building as the foundation for his science. In the *Meditations* the terms *mode* and *accident* appear to be used interchangeably. Descartes is of course thinking, in general, in substance-accident terms. During the period when he is receiving objections and replying to them, however, he is also reading Scholastic authors, especially Eustace of St. Paul, in preparation for writing the *Principles,* which will solidify his metaphysics and proceed to elaborate the physics built on its base.[12] He comes to recognize more clearly (and distinctly) just how it is he differs from the School and the distinction between mode and (real) accident becomes important to him as he is both replying to successive objections and writing the *Principles.*

Once one looks at the *Objections* and *Replies* in this context, moreover, they themselves take on the guise of a developing conversation. Thus in the Fourth Replies, in answer to his own reformulation of Arnauld's objection about sensible qualities, he begs off a confrontation by saying: "I think I can easily get round this objection if I say that I have never denied that there are real accidents. It is true that in the *Dioptrics* and the *Meteors* I did not make use of such qualities in order to explain the matters which I was dealing with, but in the *Meteors,* p. 164, I expressly said that I was not denying their existence." [13] In the Sixth Replies, however, he speaks of "the need to explode" the reality of accidents *(accidentium . . . realitatem explodendam)* and explicitly ties the conception of "real accidents" to a mistaken theory of sense perception.[14] In effect, if all perception is of surfaces, there is no need in the explanation of our contact with nature of anything but figure and other such "extensive" qualifica-

10. Disputation 7, sec. 1, par. 17, in Suárez 1982, 231–32.
11. See Garber 1986; and Beyssade 1994.
12. Ariew 1992b.
13. AT 7:248; CSM 2:173 (modified).
14. AT 7:434.

tions, which, as every one agrees, are merely modal. And he supports this position by a very Aristotelian argument that whatever is not a substance is an accident, and a "real" accident would really be another substance—hence, as in the letter to Mersenne of 1643, one of those "little souls" added to the body.[15] In short, by eliminating "accidents" (which some might think real, hence quasi-substantial) and keeping only modes, we return, paradoxically, to a much purer Aristotelian conception, where there are substances and accidents and none of those compromising in-betweens—but paradoxically also non-Aristotelian insofar as modes are all expressions of the essence of some substance, as most "accidents" were not.

What, if anything, is signified by the change in Descartes' attitude in his Fourth and Sixth Replies? It may be that in the Fourth he is replying less dogmatically because he senses in Arnauld, "his reader," the critic who enters heart and soul into his enterprise and wants, if he can—and as he will proceed for more than fifty years to do—to join it. But it may also be the case that in the period that elapses between the Fourth and Sixth Objections and Replies—with the searing experience of the Fifth in the interval, and with the work he is doing toward the formulation of the *Principles* into the bargain—Descartes may have solidified his rejection of real qualities, that is, of accidents other than modes, so that the reality of accidents is now a doctrine, not only not to be made use of, but positively to be "exploded." I do not want to make a decision on this (so far) purely speculative question, but to point out that on either reading, the *Meditations* in their original published form, complete with the *Objections* and *Replies,* clearly do constitute, as Marion argues, a complex, sometimes overt, sometimes hidden, conversation of Descartes with his contemporaries. It is a conversation conducted by a man with a vision, a vision he is still working to elaborate. As he hopes, the metaphysics, physics, and, eventually, the medicine, mechanics, and morality he is developing will serve humanity ever after as the one body of knowledge—of course alongside revealed truth—by which we will be able to regulate our lives. In this endeavor he speaks, with differing degrees of accuracy or interest, the language of his contemporaries, sometimes seeking to refine or modify, sometimes to refute their views. But he is always a speaker in and for his time, in dialogue with others of that time.

I have been dwelling on one example in order to illustrate the kind of analysis permitted by attention to the *Objections* and *Replies* as an integral part of a whole broader than the six meditations themselves. There

15. AT 3:648.

are of course other cases just as appropriate to study, and not only those included in this volume. Beyssade, in the address I have already referred to, lists a number of such instances. Marion, in particular, contributed to the *Replies* conference a detailed study of the vexed conception of *causa sui*—one of the conspicuous cases of a concept elaborated in the course of the *Replies,* and of a concept subject to intense debate before and in Descartes' time.[16] But I trust that I have adduced enough evidence to make the general point clear.

Cartesian Dialogue and the History of Philosophy

To see Descartes in conversation with his contemporaries, and, indeed, his immediate, or less immediate, predecessors, is to alter radically the approach that focuses wholly on the text itself, an approach that has been dominant, with few exceptions, at least in English-language Cartesian scholarship. I want to suggest, in conclusion, that this dialogic reading of Cartesian texts can fruitfully be extended to the works of other thinkers, and that it furnishes a useful model not only for the history of philosophy but for philosophy itself.

For one thing, *dialogue* appears to be a synonym for *dialectic,* not in the Hegelian or Marxist or Kierkegaardian sense, but in a sense that harks back to Plato, the originator of the first corpus of dialogues in our history. How do Plato's writings, which are explicitly in dialogue form, compare with the work of Descartes considered as a conversation with his critics? This may seem an odd comparison, since for much of his life Plato peopled his philosophical writings with characters who come to life (and usually were or had been alive): snub-nosed Theaetetus, who looks like Socrates, or Alcibiades, who was in love with Socrates, or the comic poet Aristophanes. Real characters and settings are quite irrelevant to the Cartesian form of intellectual exchange. It is a question of asking for objections and answering them so as to get nearer to the right answer—or of persuading others that one has had the right answer all along. But, it will be objected, isn't Plato the man who of all others thought he had right answers? Weren't there those Forms laid up in Heaven that guaranteed the truth for the few who struggled out beyond the Cave? Descartes, in contrast, has to trust to God's veracity to guarantee that his clear and distinct ideas were true. And far from being an intellectual elitist, like Plato, he believed (or so he said) that if only people would follow the order of reasons he was prescribing, they could all work forever after on the secure

16. Marion 1994.

foundation he was offering them (again, of course, given God's goodness and veracity).

There is something wrong with that picture, however. Plato never believed he had, or could have, "the right answer" in any explicit formulation. One must find the right people to talk with, seek with them for an answer, which as answer is always tentative, and perhaps at last if one is fortunate one will come, in "a flash of fire," to see with the mind's eye a truth beyond language. That is why Plato explicitly writes dialogues, all the while suspecting the adequacy of writing, dialogues in many of which the setting and the character of the speakers play an all-too-human part. That is why it is correct to say that the first great philosophical writer in our tradition never wrote down his philosophy. Descartes, on the other hand, however much his thought developed from the 1620s (and the *Rules*) through the *Discourse* and *Meditations* to the *Principles*, and however much objections to his arguments and his replies to them contributed to that development—Descartes had a vision of a unified science that, chiefly on his initiative, was to be achieved and stated in his lifetime as a possession forever. Granted, both thinkers, in contrast to Plato's rebellious student Aristotle, who loved to compartmentalize disciplines, were looking for something that is everywhere one. Perhaps one could say that Descartes was seeking a unity of knowledge, while with Plato it was rather one wisdom of which he was in search.

My comparison is a lame one at best, however, since after all Plato did write dialogues, and for good reasons, while in his published work Descartes, for equally good reasons, did not. Descartes, while welcoming Arnauld's objections, and replying with due courtesy to those presented by Caterus or Mersenne, becomes extremely irritated by the objections of Hobbes, Gassendi, or Bourdin.[17] It seems they ought to be able to understand what is quite evident, or is in the course of being made more evident as the "conversation" develops. In the Platonic dialogues, of course, the interlocutors of Socrates sometimes lose their tempers too, but Plato is quite consciously, sometimes ironically, showing us in these scenes the limitations of human intellectual exchange, limitations of which the architect of a once-for-all unified science must necessarily have been unaware, so that he responded with impatient irritation when he encountered them.

It may be more appropriate to compare Descartes with other thinkers who, though not, like Plato, writing in dialogue form, were nevertheless in some sense engaged in dialogues with their predecessors and contem-

17. See the essays in this volume by Sorrell, Curley, Osler, Lennon, and Ariew.

poraries. In the *Objections* and *Replies,* of course, this is literally the case, and Marion has argued that the *Meditations* themselves show their author replying to objections to his earlier work.[18] Yet even where there is no such development of explicit, or at least identifiable, objections and responses, there is an important sense in which other thinkers too are in conversation with their contemporaries or their predecessors. Wittgenstein's private language argument holds for the practice of philosophy as well as, perhaps even more emphatically than, for everyday concerns. Hume, for example, spending three years at Le Flèche writing his *Treatise,* begins with the language of Locke and the new way of ideas, and in his investigation into the origin of the idea of cause, replies explicitly to arguments of Locke, Clarke, and others.[19] Granted, his work, as he lamented, fell dead born from the press. His attempt, in the light of this nonresponse, to modify his argument in the direction of a more "literary" presentation may be taken as a form of reply to the objections of silence. And the *Dialogues on Natural Religion* (this time explicitly in dialogue form, of a Ciceronian variety), though published posthumously, present a sustained reply to and development of positions current in his time. His letters show him trying to enter into those perspectives so as to develop their presentation convincingly, particularly in the case of Cleanthes, the character least credible though probably most common in Hume's Edinburgh.[20]

To take one more example: Marion contrasts the work of Spinoza, as a "solipsistic" thinker, with the dialogic style of Cartesian thought. And it is true that the *Ethics* looms as a transcendent text, worthy (as the *Meditations* have hitherto appeared to many) of prolonged and detailed study within the four corners of the document. Yet Spinoza began his philosophical enterprise, as far as we have evidence of its beginnings, by entering into conversation with the thought of Descartes, in the *Principles of Cartesian Philosophy* and the *Cogitata metaphysica.* He was also, as the *Correspondence* makes clear, engaged in discussion and debate with his contemporaries. For his relation to traditional thinkers, consider, even

18. See Marion's essay in this volume.

19. The Cartesian distinction between substance and mode, oddly transmogrified by Locke, is given a bow at the opening of the *Treatise,* and Locke's added category of relation becomes central to Hume's investigation. And "ideas" in their Lockean, no longer Cartesian, version are split into impressions and ideas, now but their fainter copies. Hume is carrying through to its logical conclusion Locke's (mis)understanding of Cartesian method.

20. On March 10, 1751, Hume wrote to Gilbert Elliott of Minto: "You wou'd perceive by the Sample I have given you, that I make Cleanthes the Hero of the Dialogue. Whatever you can think of, to strengthen that side of the argument, will be most acceptable to me" (Hume 1932, 153–54).

apart from the *Tractatus Theologico-Politicus,* the appendix to *Ethics,* part 1, with its challenge to traditional religious ontologies, whether Protestant, Catholic, or Jewish. This is no solipsist, but a participant in a complex nexus of communities.[21] And above all—or beneath it—Spinoza was seeking to guide his fellow men, not only himself, but all who would listen, "by the hand to the highest human happiness." True, his major work, like Hume's *Dialogues,* was only posthumously published. But even though it was officially decried as "horrid atheism," there were always here and there ears to listen, so that dialogue with Spinoza may be said to continue to this day.[22]

Starting from the conclusion of Marion's essay, and his placement of Cartesian thought "in the . . . space of dialogue," I have extended his remark, far beyond its seeming intent, into the space of metaphor. But I do want to suggest, as a conclusion to this volume, that a contextual approach, not only to Descartes, but to any or all of our traditional texts, can shed important new light, not only on those texts themselves but on the history of philosophy as such and on the nature of philosophy as (a) history.

First, for the practice of the history of philosophy, it is necessary, as in any conversation, to enter into the universe of discourse of one's interlocutor. As Plato understood, this is seldom easy, and indeed, it is sometimes impossible. But even to attempt it, we have to set aside some of our own immediate philosophical concerns and learn to enter in imagination into the language and beliefs of the participants in the case in question. No text written earlier than yesterday can help us do this without some external aid, the kind of aid the essays in this volume are intended to provide. If we want to engage in a dialogue with Descartes or Hume or Spinoza, for example, we must try to understand what the philosopher in question was saying—and what he may be saying to us if we can succeed in our task of imaginative projection—not through a given text itself (let alone a rendering into another language of such a text), but in terms of the dialogue of the philosopher in question with his contemporaries and/or predecessors in their time(s) and place(s).

Finally, if I may generalize still further (but this is not a "hasty generalization"), the history of philosophy understood in contextual perspective

21. Although Spinoza had rejected, and been rejected by, the Jewish community in which he had his roots, he was actively engaged in the development of biblical criticism—in which he now appears a pioneer. He actively participated in the development of the mechanical natural philosophy, particularly in its Cartesian form. And, friend of statesman Jan De Witt, he was also concerned with the political problems of the Netherlands.

22. See Deleuze 1981.

appears to be continuous with philosophy itself, or rather with Western philosophy, the tradition in which we take, or find, our place. If we see Descartes in dialogue with the School or with others who oppose the School, though on very different grounds from his, we see Locke in dialogue with Descartes, as well as with the honorable Mr. Boyle or the incomparable Newton. We see Leibniz in dialogue with Locke, as well as with Descartes—via Arnauld—Kant answering Hume as well as Leibniz-Wolff, and so on and on. And in the last analysis we see philosophy itself as a conversation stretching from Elea to Oxford or from Meletus to San Diego. Some episodes are richer, some poorer, some more fruitful, some more arid than others. But it is in our place in that history that we develop our own philosophical reflections. Thus the effort to understand the thinkers of the past in their intellectual and cultural contexts should help us to enrich and enlarge the perspective in which we ourselves venture to philosophize.

BIBLIOGRAPHY

Adam, Charles. 1910. *Vie et oeuvres de Descartes.* Paris: Cerf.

Adams, Marilyn McCord. 1970. "Intuitive Cognition, Certainty, and Scepticism in William of Ockham." *Traditio* 26:389–98.

Alféry, Pierre. 1989. *Guillaume d'Occam: Le singulier.* Paris: Editions de Minuit.

Alquié, Ferdinand. 1974. *Le cartésianisme de Malebranche.* Paris: Vrin.

Al-Qur'an: A Contemporary Translation. 1990. Translated by Ahmed Ali. Princeton, N.J: Princeton University Press.

Andreas, Valerius. 1739. *Bibliotheca belgica.* 2 vols. In Jean-François Foppeu, *Bibliotheca belgica.* Brussels.

Angelis, Enrico de. 1964. *Il metodo geometrico nella filosofia del Seicento.* Pisa: Istituto di Filosofia.

Argentré, Duplessis d'. 1736. *Collectio judiciorum de novis erroribus tomus tertium.* Paris.

Ariew, Roger. 1987. "The Infinite in Descartes' Conversation with Burman." *Archiv für Geschichte der Philosophie* 69:140–63.

———. 1992a. "Bernier et les doctrines gassendistes et cartésiennes de l'espace." *Corpus* 20/21:155–70.

———. 1992b. "Descartes and Scholasticism: The Intellectual Background to Descartes' Thought." In *Cambridge Companion to Descartes,* edited by J. Cottingham, 58–90. Cambridge: Cambridge University Press.

———. 1994. "Sur les septièmes réponses." In *Descartes: Objecter et répondre,* edited by J.-M. Beyssade and J.-L. Marion, 123–40. Paris: P.U.F.

———. Forthcoming. "Descartes and the Jesuits." In *The Jesuits and the Scientific Revolution,* edited by M. Feingold. Princeton, N.J.: Princeton University Press.

Aristotle. 1941. *The Basic Works.* Edited by R. McKeon. New York: Random House.

———. 1984. *The Complete Works of Aristotle.* Edited by Jonathan Barnes. 2 vols. Princeton, N.J.: Princeton University Press.

Armogathe, Jean-Robert. 1977. *Theologia cartesiana: L'explication physique de l'Eucharistie chez Descartes et Dom Desgabets.* International Archives of the History of Ideas 84. The Hague: Martinus Nijhoff.

———. 1987. "L'arc-en-ciel dans les *Météores*." In *Le discours et sa méthode,* edited by M. Grimaldi and J.-L. Marion, 145–62. Paris: P.U.F.

———. 1994. "L'approbation des *Meditationes* par la Faculté de Théologie de Paris (1641)." *Bulletin Cartésien* 21:1–3. In *Archives de Philosophie 57.*

Arnauld, Antoine. 1662. *La logique, ou L'art de penser.* Edited by Bruno Baron von Freytag Löringhoff and Herbert E. Brekle. Vol. 1. Paris. (Facsimile reprint, Stuttgart: Friedrich Frommann, 1965.)

———. 1775–83. *Oeuvres.* 43 vols. Paris: S. d'Arnay.

Atheniensis, Philalethus Eleuterius. 1648. *Specimen tum inscitia tum malitia detectae in calumniis et mendaciis partim Stevartii . . . partim Revii.* Dicaeopolis.

Aubrey, John. [1680] 1975. *Brief Lives.* Edited by R. Barber. London: Folio Society.

Babin, François. 1679. *Journal ou relation fidele de tout ce qui s'est passé dans l'université d'Angers au sujet de la philosophie de Des Carthes en l'execution des ordres du Roy pendant les années 1675, 1676, 1677, et 1678.* Angers.

Baillet, Adrien. 1691. *La vie de Monsieur Des-Cartes.* 2 vols. Paris.

Barrow, Isaac. 1860. *The Mathematical Works.* Edited by William Whewell. Cambridge. (Facsimile reprint, Hildesheim: G. Olms, 1973.)

Baudry, Léon. 1950. *Guillaume d'Occam: Sa vie, ses oeuvres, ses idées sociales et politiques.* Paris: Vrin.

———. 1958. *Lexique philosophique de Guillaume d'Ockham.* Paris: Lethellieux.

Baumgartner, Frederic J. 1987. "Sunspots or Sun's Planets: Jean Tarde and the Sunspot Controversy of the Early Seventeenth Century." *Journal for the History of Astronomy* 18:44–54.

Bayle, Pierre. 1684. *Recueil de quelques pièces curieuses concernant la philosophie de Monsieur Descartes.* Amsterdam.

———. 1720. *Dictionnaire historique et critique.* 4 vols. Rotterdam.

Beaulieu, Armand. 1990. "Les relations de Hobbes et de Mersenne." In *Thomas Hobbes, philosophie première, thèorie de la science, et politique,* edited by Y.-C. Zarka and J. Bernhardt, 81–90. Paris: P.U.F.

Besoigne, J. 1752. *Histoire de l'abbaye de Port-Royal.* Paris.

Beyssade, J.-M. 1994. "Méditer, objecter, répondre." In *Descartes: Objecter et répondre,* edited by J.-M. Beyssade and J.-L. Marion, 21–40. Paris: P.U.F.

Bibliotheca Catholica Neerlandica Impressa, 1500–1727. 1954. The Hague: Nijhoff.

Bloch, O. R. 1966. "Gassendi critique de Descartes." *Revue Philosophique* 156: 217–36.

———. 1971. *La philosophie de Gassendi: Nominalisme, matérialisme et métaphysique.* The Hague: Martinus Nijhoff.

Bos, H. J. M. 1981. "On the Representation of Curves in Descartes' *Géométrie.*" *Archive for History of Exact Sciences* 24:295–338.

Bougerel, J. 1737. *Vie de Pierre Gassendi.* Paris.

Bouillier, Francisque. 1868. *Histoire de la philosophie cartésienne.* 3d ed. Vol. 1. Paris: Delagrave.

Bourdin, Pierre. 1639. *Prima geometriae elementa.* Paris.

———. 1640. *Geometria, nova methodo*. Paris.

———. 1643. *L'introduction à la mathématique*. Paris.

———. 1646. *Sol flamma sive tractatus de sole, ut flamma est, eiusque pabulo sol exurens montes, et radios igneos exsufflans Eccles. 43. Aphorismi analogici parvi mundi ad magnum ad parvuum*. Paris.

———. 1655a. *L'architecture militaire, ou L'art de fortifier les places regulières et irregulières*. Paris.

———. 1655b. *Le dessein, ou La perspective militaire*. Paris.

———. 1661. *Le cours de mathématique*. 3d ed. Paris.

Boyle, Robert. 1991. *Selected Philosophical Papers of Robert Boyle*. Edited by S. Stewart. Indianapolis: Hackett.

Bracken, H. M. 1964. "Some Problems of Substance among the Cartesians." *American Philosophical Quarterly* 1:129–37.

Brandt, F. 1928. *Hobbes' Mechanical Conception of Nature*. London: Hachette.

Brehier, Emile. 1967. "The Creation of the Eternal Truths in Descartes's System." In *Descartes*, edited by W. Doney, 192–208. Garden City, N.Y.: Anchor Books.

Brockliss, L. W. B. 1987. *French Higher Education in the Seventeenth and Eighteenth Centuries: A Cultural History*. Oxford: Clarendon Press.

Bruinvisch, C. W. 1905. "Admissiën van Priesters te Alkmaar, 1641–1727." *Bijdragen voor de geschiedenis van het Bisdom Haarlem* 29:252–61.

Campanus de Novarra. 1482. *Opus elementorum euclidis megarensis in geometriam artem*. Venice.

Carabellese, Pantaleo. 1946. *Le obbiezioni al cartesianesimo*. Biblioteca di Cultura Moderna. 3 vols. Messina: D'Anna.

Carraud, Vincent. 1992. *Pascal et la philosophie*. Paris: P.U.F.

Casilius, Antonius. 1643. *Introductio in Aristotelis logicam et reliquas disciplinas*. Rome.

Caton, Hiram. 1975. "Will and Reason in Descartes's Theory of Error." *Journal of Philosophy* 72:87–104.

Cicero. 1933. *De natura deorum*. Translated by H. Rackham. Cambridge: Harvard University Press.

Clauberg, Joannis. [1691]. 1968. *De cognitione Dei et nostri*. In *Opera omnia philosophica*. (Hildesheim: Olms.)

Clavius, Christopher. 1611–12. *Opera mathematica*. 4 vols. Moguntiae: Eltz.

Commandino, Federigo. 1619. *Degli Elementi d'Euclide libri quindici*. Pesaro.

———. 1620. *Elementorum Euclidis libri tredecim*. London.

Cordemoy, Géraud de. 1968. *Six discours sur la distinction et l'union du corps et de l'âme*. In *Oeuvres philosophiques*, edited by P. Clair and F. Girbal, 87–189. Paris: P.U.F.

Cornford, Francis MacDonald. [1937] 1957. *Plato's Cosmology: The "Timaeus" of Plato*. Indianapolis: Bobbs-Merrill.

Costabel, Pierre. 1982. "La controverse Descartes-Roberval au sujet du centre d'oscillation." In *Démarches originales de Descartes savant*, 167–81. Paris: Vrin.

Courtenay, William J. 1985. "The Dialectic of Omnipotence." In *Divine Omnipotence and Omniscience in Medieval Philosophy*, edited by Tamar Rudavsky, 243–69. Dordrecht: Reidel.

Crombie, Alistair C. 1977. "Mathematics and Platonism in the Sixteenth-Century Italian Universities and in Jesuit Educational Policy." In *Prismata: Naturwissenschaftsgeschichtliche Studien Festschrift für Willy Hartner*, edited by Y. Maeyama and W. G. Saltzer, 63–94. Wiesbaden: Franz Steiner.

Curley, Edwin. 1978. *Descartes against the Skeptics*. Cambridge: Harvard University Press.

———. 1984. "Descartes on the Creation of the Eternal Truths." *Philosophical Review* 93:569–97.

———. 1986. "Analysis in the *Meditations:* The Quest for Clear and Distinct Ideas." In *Essays on Descartes' Meditations*, edited by A. O. Rorty, 153–76. Berkeley: University of California Press.

———. 1992. "'I Durst Not Write So Boldly'; or, How to Read Hobbes' Theological-Political Treatise." In *Hobbes e Spinoza*, edited by Daniela Bostrenghi, 497–593. Naples: Bibliopolis.

Dalbiez, R. 1929. "Les sources scolastiques de la théorie cartésienne de l'être objectif (à propos du Descartes de M. Gilson)." *Revue d'Histoire de la Philosophie* 3:464–72.

Dear, Peter. 1987. "Jesuit Mathematical Science and the Reconstitution of Experience in the Early Seventeenth Century." *Studies in History and Philosophy of Science* 18:133–75.

———. 1988. *Mersenne and the Learning of the Schools*. Ithaca, N.Y.: Cornell University Press.

———. 1995. *Disciplined Experience: The Mathematical Way in the Scientific Revolution*. Chicago: University of Chicago Press.

Deleuze, Gilles. 1981. *Spinoza: Philosophie pratique*. Paris: Editions de Minuit.

Descartes, René. [1911] 1967. *Philosophical Works*. Translated by E. S. Haldane and G. R. T. Ross. 2 vols. Cambridge: Cambridge University Press.

———. 1936–63. *Correspondence de Descartes*. Edited by Charles Adam and Géraud Milhaud. 8 vols. Vols. 1–2, Paris: Alcan. Vols. 3–8, Paris: P.U.F.

———. 1963–73. *Oeuvres philosophiques*. Edited and translated by Ferdinand Alquié. 3 vols. Paris: Garnier.

———. 1964–76. *Oeuvres de Descartes*. Edited by Charles Adam and Paul Tannery. 11 vols. Rev. ed. Paris: Vrin.

———. 1965. *Discourse on Method, Optics, Geometry, and Meteorology*. Translated by P. J. Olscamp. Indianapolis: Bobbs-Merrill.

———. 1970. *Descartes: Philosophical Letters*. Translated and edited by Anthony Kenny. Oxford: Clarendon Press.

———. 1976. *Discours de la Méthode*. Text with commentary by Etienne Gilson. Paris: Vrin.

———. 1979. *Descartes*. Edited by J.-M. Beyssade and M. Beyssade. Paris: P.U.F.

———. 1985. *The Philosophical Writings of Descartes.* Translated by J. Cottingham, R. Stoothoff, and D. Murdoch. 2 vols. Cambridge: Cambridge University Press.

Descartes, René, and Martin Schoock. 1988. *Le querelle d'Utrecht.* Edited and translated by Theo Verbeek. Paris: Les Impressions Nouvelles.

Desharnais, Richard P. 1966. "The History of the Distinction between God's Absolute and Ordained Power and Its Influence on Martin Luther." Ph.D. diss., Catholic University of America.

Dibon, Paul. 1990. "La réception du *Discours de la méthode* dans les Provinces Unies." In *Descartes: Il metodo e i saggi,* edited by G. Belgioioso et al., 625–50. Rome: Instituto della Enciclopedia Italiana.

Doney, Willis. 1978. "The Geometrical Presentation of Descartes's A Priori Proof." In *Descartes: Critical and Interpretive Essays,* edited by Michael Hooker, 1–25. Baltimore: Johns Hopkins University Press.

Duhamel, Jean. 1692. *Reflexions critiques sur le système cartésien de la philosophie de mr. Régis.* Paris.

Euclid. 1956. *The Thirteen Books of Euclid's Elements.* Edited and translated by Thomas L. Heath. 2 vols. New York: Dover.

Eusebius. 1965. *The History of the Church from Christ to Constantine.* Translated by G. A. Williamson. New York: Dorset Press.

Eustachius a Sancto Paulo. 1620. *Summa philosophiae quadripartita.* Cologne.

[Fabri, Honoré]. 1646. *Philosophiae tomus primus: Qui complectitur scientiarum methodum sex libris explicatam: Logicam analyticam, duodecim libris demonstratam, & aliquot controversias logicas, breviter disputatas. Auctore Petro Mosnerio Doctore Medico. Cuncta excerpta ex praelectionibus R. P. Hon. Fabry.* . . . Lyons.

Foster, M. B. 1936. "Christian Theology and Modern Science of Nature II." *Mind* 45: 1–28.

Foucher, Simon. [1675] 1969. *Critique de la recherche de la vérité.* With an introduction by R. A. Watson. New York: Johnson Reprint.

Frankfurt, Harry. 1977. "Descartes and the Creation of Eternal Truths." *Philosophical Review* 86:40–41.

Funkenstein, Amos. 1986. *Theology and the Scientific Imagination from the Middle Ages to the Seventeenth Century.* Princeton, N.J.: Princeton University Press.

Galilei, Galileo. 1953. *Dialogue Concerning the Two Chief World Systems, Ptolemaic and Copernican.* Translated by Stillman Drake. Berkeley: University of California Press.

Galluzzi, Paolo. 1973. "Il 'Platonismo' del tardo Cinquecento e la filosofia di Galileo." In *Ricerche sulla cultura dell'Italia moderna,* edited by Paola Zambelli, 37–79. Bari: Laterza.

Garber, Daniel. 1986. "*Semel in vita:* The Scientific Background to Descartes' *Meditations*." *Essays in Descartes' "Meditations,"* edited by A. O. Rorty, 81–116. Berkeley: University of California Press.

————. 1987. "How God Causes Motion: Descartes, Divine Sustenance, and Occasionalism." *Journal of Philosophy* 84:567–80.

————. 1988. "Descartes, the Aristotelians, and the Revolution That Didn't Happen in 1637." *Monist* 71:471–86.

————. 1992. *Descartes' Metaphysical Physics.* Chicago: University of Chicago Press.

————. 1993. "Descartes and Occasionalism." In *Causation in Early Modern Philosophy: Cartesianism, Occasionalism, and Preestablished Harmony,* edited by S. Nadler, 9–26. University Park: Pennsylvania State University Press.

Garber, Daniel, and Lesley Cohen. 1982. "A Point of Order: Analysis, Synthesis, and Descartes' *Principles.*" *Archiv für Geschichte der Philosophie* 64:136–47.

Gassendi, Pierre. [1658] 1964. *Opera omnia.* 6 vols. Stuttgart: Friedrich Frommann.

————. 1962. *Disquisitio metaphysica seu dubitationes et instantiae adversus Renati Cartesii metaphysicam et responsa.* Edited and translated into French by Bernard Rochot. Paris: Vrin.

Gaukroger, Stephen. 1992. "The Nature of Abstract Reasoning: Philosophical Aspects of Descartes' Work in Algebra." In *The Cambridge Companion to Descartes,* edited by J. Cottingham, 91–114. Cambridge: Cambridge University Press.

Geach, Peter. 1972. "God's Relation to the World." In *Logic Matters,* 318–27. Berkeley: University of California Press.

Geulincx, Arnold. [1691] 1892. *Metaphysica vera.* In *Opera philosophica,* edited by J. P. N. Land, 3:139–98. 3 vols. The Hague.

Gilson, Etienne. 1913. *La liberté chez Descartes et la théologie.* Paris: Vrin.

————. 1976. *Commentaire du "Discours de la méthode."* 5th ed. Paris: Vrin.

————. 1979. *Index scolastico-cartésien.* 2d ed. Paris: Vrin.

————. 1984. *Etudes sur le rôle de la pensée médiévale dans la formation du système cartésien.* 5th ed. Paris: Vrin.

Gouhier, Henri. 1977. *Fénelon philosophe.* Paris: Vrin.

————. 1978. *Cartésianisme et augustinisme au XVIIe siècle.* Paris: Vrin.

————. 1987a. *L'anti-humanisme au XVII e siècle.* Paris: Vrin.

————. 1987b. *La pensée métaphysique de Descartes.* 4th ed. Paris: Vrin.

Goujet, Claude-Pierre. 1758. *Mémoire historique et littéraire sur le Collège Royal de France.* 3 vol. Paris.

Graaf, J. J. 1887. "Uit de Akten van het Haarlemsch Kapittel." *Bijdragen voor de geschiedenis van het Bisdom Haarlem* 14:7.

Graves, Robert. 1955. *The Greek Myths.* 2 vols. Harmondsworth, Eng.: Penguin.

Grene, Marjorie. 1985. *Descartes.* Minneapolis: University of Minnesota Press.

————. 1991. *Descartes among the Scholastics.* Milwaukee: Marquette University Press.

Gueroult, Martial. [1968] 1984–85. *Descartes According to the Order of Reasons.* Translated by R. Ariew. 2 vols. Minneapolis: University of Minnesota Press.

Hacking, Ian. 1975. *The Emergence of Probability*. Cambridge: Cambridge University Press.

Hatfield, Gary. 1986. "The Senses and the Fleshless Eye: The Meditations as Cognitive Exercises." In *Essays on Descartes' "Meditations,"* edited by A. O. Rorty, Berkeley: University of California Press.

Hervey, Helen. 1952. "Hobbes and Descartes in the Light of Some Unpublished Letters of the Correspondence between Sir Charles Cavendish and Dr. John Pell." *Osiris* 10: 67–90.

Heusden, Hugo Franciscus van. 1714. *Batavia Sacra sive res gestae apostolicorum virorum qui fidem Bataviae primi imtulerunt*. 2 vols. Brussels: Foppens.

Hobbes, Thomas. 1839–45. *Opera latina*. Edited by William Molesworth. 5 vols. London.

———. 1969. *The Elements of Law, Natural and Politic*. Edited by F. Tonnies. London: Frank Cass.

———. 1994. *Leviathan*. Edited by Edwin Curley, with a translation of variant passages from the Latin edition of 1668. Indianapolis: Hackett.

Huet, Pierre Daniel. 1689. *Censura philosophiae cartesianae*. Paris.

Hume, David. 1932. *The Letters of David Hume*. Edited by J. Y. T. Grieg. Vol. 1. Oxford: Clarendon Press.

Iwanicki, Joseph. 1936. *Morin et les démonstrations mathématiques de l'existence de Dieu*. Paris: Vrin.

Jardine, Nicholas. 1976. "Galileo's Road to Truth and the Demonstrative Regress." *Studies in History and Philosophy of Science* 7:277–318.

———. 1988. "Epistemology of the Sciences." In *The Cambridge History of Renaissance Philosophy*, edited by Charles B. Schmitt, Quentin Skinner, Eckhard Kessler, and Jill Kraye, 685–71. Cambridge: Cambridge University Press.

Jolley, Nicholas. 1990. *The Light of the Soul: Theories of Ideas in Leibniz, Malebranche, and Descartes*. Oxford: Clarendon Press.

Jones, Howard. 1981. *Pierre Gassendi, 1592–1655: An Intellectual Biography*. Nieuwkoop: B. De Graaf.

Jones, P. J. 1947. "The Identity of the Author of a Hitherto Anonymous Work." *Scripta Mathematica* 13:119–20.

Knappich, Wilhelm. 1986. *Histoire de l'astrologie*. Paris: Editions du Féslin.

Knuttel, W. P. C. 1892–94. *De toestand der Nederlandsche Katholieken ten tijde der Republiek*. 2 vols. 's-Gravenhage: Martinus Nijhoff.

La Forge, Louis de. 1974. *Traité de l'espirit de l'homme*. In *Oeuvres philosophiques*, edited by P. Clair, 69–349. Paris: P.U.F.

Leibniz, G. W. 1965. *Die philosophischen Schriften von G. W. Leibniz*. Edited by C. J. Gerhardt. 7 vols. Hildesheim: G. Olms.

Lennon, Thomas M. 1974. "Occasionalism and the Cartesian Metaphysic of Motion." *Canadian Journal of Philosophy*, supplementary vol. 1 , pt. 1:29–43.

———. 1991. "The Epicurean New Way of Ideas: Gassendi, Locke, and Berkeley." In *Atoms, Pneuma, and Tranquility*, edited by Margaret J. Osler, 259–71. Cambridge: Cambridge University Press.

——. 1993. *The Battle of the Gods and Giants: The Philosophical Legacies of Descartes and Gassendi, 1655–1715.* Princeton, N.J.: Princeton University Press.

Lenoble, Robert. [1943] 1971. *Mersenne, ou La naissance du mécanisme.* Paris: Vrin.

Lenoir, Timothy. 1979. "Descartes and the Geometrization of Thought: The Methodological Foundation of Descartes' *Géométrie.*" *Historia Mathematica* 6:355–79.

Lloyd, G. E. R. 1979. *Magic, Reason, and Experience: Studies in the Origins and Development of Greek Science.* Cambridge: Cambridge University Press.

Locke, John. 1823. *The Works.* 10 vols. London: T. Tegg.

——. 1975. *An Essay Concerning Human Understanding.* Edited by P. H. Nidditch. Oxford: Clarendon Press. (Corrected 1979.)

Mackie, J. L. 1974. "Locke's Anticipation of Kripke." *Analysis* 34:177–80.

——. 1976. *Problems from Locke.* Oxford: Clarendon Press.

Maierù, Luigi. 1978. "Il quinto postulato euclideo in Cristoforo Clavio." *Physis* 20:191–212.

——. 1982. "Il quinto postulato euclideo da C. Clavio, 1589, a G. Saccheri, 1733." *Archive for History of Exact Sciences* 27:297–334.

——. 1984. "Il meraviglioso problema' in Oronce Finé, Girolamo Cardano e Jacques Peletier." *Bolletino di Storia delle Scienze Matematiche* 4:141–70.

——. 1989. "Le vicende relative al quinto postulato euclideo fra il Cinquecento e il Settecento." In *La cultura filosofica e scientifica. Tomo secondo: La storia delle scienze,* edited by C. Maccagni and P. Freguglia, 127–59. Busto Arsizio: Bramante.

——. 1990. "'. . . in Christophorum Clavium de Contactu Linearum Apologia': Considerazioni attorno alla polemica fra Peletier e Clavi circa l'angolo di contatto, 1579–1589." *Archive for History of Exact Sciences* 41:115–37.

Malebranche, Nicolas. 1958–69. *Oeuvres complètes.* Paris: Vrin.

——. 1980. *The Search after Truth,* translated by T. M. Lennon and P. J. Olscamp. *Elucidations,* translated by T. M. Lennon. *Philosophical Commentary,* by T. M. Lennon. Columbus: Ohio State University Press.

Mancosu, Paolo. 1992. "Aristotelean Logic and Euclidean Mathematics: Seventeenth-Century Developments of the Quaestio de certitudine mathematicarum." *Studies in History and Philosophy of Science* 23:241–65.

Marion, Jean-Luc. 1981. *Sur la théologie blanche de Descartes.* Paris: P.U.F.

——. 1986a. "The Essential Incoherence of Descartes' Definition of Divinity." In *Essays on Descartes's "Meditations,"* edited by A. O. Rorty, 297–335. Berkeley: University of California Press.

——. 1986b. *Sur le prisme métaphysique de Descartes.* Paris: P.U.F.

——. 1991. "Quelle est la métaphysique de la méthode? La situation métaphysique du *Discours de la méthode.*" In *Questions cartésiennes,* 37–73. Paris: P.U.F.

——. 1994. "Entre analogie et principe de raison: "La *causa sui.*" In *Descartes: Objecter et répondre,* edited by J.-M. Beyssade and J.-L. Marion, 305–36. Paris: P.U.F.

Martinet, Monette. 1986. "Jean-Baptiste Morin (1583–1656)." In *Quelques savants et amateurs de Science au XVIIe siècle: Sept notices biobibliographiques caractéristiques,* edited by P. Costabel and M. Martinet, 69–87. Paris: Société Française d'Histoire des Sciences et des Techniques and Editions Belin.

Martinich, A. P. 1992. *The Two Gods of "Leviathan."* Cambridge: Cambridge University Press.

Matricule de l'Université de Louvain. 1962. Vol. 5. Brussels: Palais des Académies.

McMullin, Ernan. 1978. *Newton on Matter and Activity.* Notre Dame, Ind.: University of Notre Dame Press.

McRae, Robert. 1972. "Innate Ideas." In *Cartesian Studies,* edited by R. J. Butler, 32–54. Oxford: Basil Blackwell.

Menn, Stephen. Forthcoming. "Suárez, Nominalism, and Modes." In *Hispanic Philosophy in the Age of Discovery.* Washington, D.C.: Catholic University of America Press.

Mersenne, Marin. 1623. *Quaestiones celeberrimae in Genesim.* Paris.

———. 1625. *La vérité des sciences.* Paris.

———. 1932–88. *Correspondence du P. Marin Mersenne, religieux minime.* Edited by C. de Waard et al. 17 vols. Vol. 1, Paris: Beau-Chesne. Vols. 2–4, Paris: P.U.F. Vols. 5–17, Paris: CNRS.

Meyjes, E. J. W. 1895. "Jacobus Revius: Zijn leven en zijn werken." Ph.D. diss., University of Utrecht.

Milton, J. R. 1987. "Induction before Hume." *British Journal for the Philosophy of Science* 38:49–74.

Molland, A. G. 1976. "Shifting the Foundations: Descartes's Transformation of Ancient Geometry." *Historia Mathematica* 3:21–49.

Monchamp, Georges. 1886. *Histoire du cartésianisme en Belgique.* Brussels: F. Hayez.

Morin, Jean-Baptiste. 1619. *Nova mundi sublunaris anatomia.* Paris.

———. 1623. *Astrologicarum domorum cabala detecta.* Paris.

———. 1624. *Réfutation des thèses erronées d'Anthoine Villon . . . & Estienne de Claues. . . .* Paris.

———. 1628. *Ad australes et boreales astrologos; Pro astrologia restituenda epistolae.* Paris.

———. 1635. *Quod Deus sit.* Paris.

———. 1651. *Defensio . . . dissertationis de Atomis et Vacuo adversus P. Gassendi Philosophiam Epicuream, contra Fr. Bernerii Anatomiam ridiculi murus.* Paris.

———. 1661. *Astrologia gallica.* The Hague.

Mous, A. Th. 1962–67. "Geschiedenis van het voormalig kapittel van de kathedrale kerk van Sint-Bavo te Haarlem, 1561–1616." *Archief voor de geschiedenis van de Katholieke Kerk in Nederland* 4:75–123, 125–84, 295–337; 6:257–90; 8:257–86; 9:276–317.

Nadler, Steven. 1989. *Arnauld and the Cartesian Philosophy of Ideas*. Princeton, N.J.: Princeton University Press.

———. 1992. *Malebranche and Ideas*. Oxford: Oxford University Press.

———. 1994. "Descartes and Occasional Causation." *British Journal for the History of Philosophy*. 2:35–54.

———. Forthcoming. "Occasionalism and the Mind-Body Problem." *Oxford Studies in the History of Philosophy 2*.

Naux, Charles. 1983. "Le père Christophore Clavius, 1537–1612: Sa vie et son oeuvre." *Revue des Questions Scientifiques* 154:55–67, 181–93, 325–47.

Ndiaye, Aloyse Raymond. 1991. *La philosophie d'Antoine Arnauld*. Paris: Vrin.

Neveu, Bruno. 1969. *Sebastian Joseph du Cambout de Pontchateaus, 1634–1690, et ses missions à Rome d'après sa correspondance et des documents inédits*. Paris: E. de Boccard.

Norris, John. 1690. *Cursory Reflections upon . . . "An Essay Concerning Human Understanding."* London.

Oakley, Francis. 1984. *Omnipotence, Covenant, and Order: An Excursion in the History of Ideas from Abelard to Leibniz*. Ithaca, N.Y.: Cornell University Press.

Oldenburg, Henry. 1966. *The Correspondence of Henry Oldenburg*. Edited by A. Rupert Hall and Marie Boas Hall. Vol. 2. Madison: University of Wisconsin Press.

Orcibal, J. 1989. *Jansenius d'Ypres, 1585–1638*. Paris: Etudes Augustiniennes.

Osler, Margaret J. 1994. *Divine Will and the Mechanical Philosophy: Gassendi and Descartes on Contingency and Necessity in the Created World*. Cambridge: Cambridge University Press.

Pascal, Blaise. 1963. "De l'esprit géométrique et de l'art de persuader," in *Pascal: Oeuvres complètes*, edited by Louis Lafuma, 348–59. Paris: Editions du Seuil.

Peletarius, Jacobus. 1557. *In Euclidis Elementa Geometrica demonstrationum libri sex*. Lyons.

———. 1579. *In Christophorum Clavium, de contactu linearum, apologia*. Paris.

Pérez-Ramos, Antonio. 1988. *Francis Bacon's Idea of Science and the Maker's Knowledge Tradition*. Oxford: Clarendon Press.

Pintard, R. 1943. *Le libertinage érudit dans la première moitié du XVIIe siècle*. 2 vols. Paris: Boivin.

Poelhekke, J. J. 1964. *Het geval Zijdewind*. Amsterdam: Noord-Hollandse Uitgeversmaatschappij.

Popkin, Richard H. 1979. *The History of Scepticism from Erasmus to Spinoza*. Berkeley: University of California Press.

Proclus. 1970. *A Commentary on the First Book of Euclid's Elements*. Translated by Glenn R. Morrow. Princeton, N.J.: Princeton University Press.

Prost, J. 1907. *Essai sur l'atomisme et l'occasionalisme dans la philosophie cartésienne*. Paris: Henry Paulin.

Quesnel, P. 1697. *Histoire de la vie et des ouvrages de M. Arnauld*. Liège.

Rapin, René. 1725. *Reflexions sur la philosophie.* In *Oeuvres.* Paris.

Reif, Sister Mary Richard. 1962. "Natural Philosophy in Some Early Seventeenth-Century Scholastic Textbooks." Ph.D. diss., St. Louis University.

Reneri, Henricus. N.d. Correspondence with De Wilhem. Leiden University Library. MS Dept., BPL 293A.

Revius, Jacobus. 1642–46. *Disputationes in universam theologiam.* Leiden: J. N. à Dorp.

———. 1646. *Suárez repurgatus.* Leiden.

———. 1646–53. *Analectorum theologicorum disputationes CCCXXXI.* Ludg. Bat.

———. [1930–35] 1976. *Over-Ysselsche Sangen en Dichten.* Ed. W. A. P. Smit. 2 vols. Amsterdam: Uitg. Mij. Holland Reprint. Utrecht: HES Publishers.

———. 1648. *Methodi cartesiani consideratio theologica.* Lugd. Bat.

———. 1650. *Statera philosophiæ cartesianae.* Lugd. Bat.

———. 1968. *Jacob Revius: A Dutch Metaphysical Poet.* Selected and translated by Henrietta ten Harmsel. Detroit: Wayne State University Press.

Rijkenberg, E. H. 1896. "De geschiedenis en de reliquie van het Mirakel van het Heilig Bloed te Alkmaar." *Bijdragen voor de geschiedenis van het Bisdom Haarlem* 21:321–409.

Risse, Wilhelm. 1964. *Die Logik der Neuzeit, Bd. 1, 1500–1640.* Stuttgart: Friedrich Frommann.

———. 1970. *Die Logik der Neuzeit, Bd. 2, 1640–1780.* Stuttgart: Friedrich Frommann.

Rochemonteix, Camille de. 1889. *Un Collège de Jèsuites aux 17e et 18e siècles.* 4 vols. Le Mans.

Rodis-Lewis, G. 1973. "Cartésianisme et jansénisme, Arnauld et Nicole." In "Descartes—cartésiens et anticartésiens français." In *Histoire de la philosophie.* II. Paris: Encyclopédie de la Pléiade.

———. 1986. "On the Complementarity of Meditations III and V: From the 'General Rule' of Evidence to 'Certain Science.'" In *Essays on Descartes's "Meditations,"* edited by A. O. Rorty, 271–91. Berkeley: University of California Press.

———. 1988. "L'acceuil fait aux *Méteores.*" In *Problématique et réception du Discours et des Essais,* edited by H. Mechoulan, 99–108. Paris: Vrin.

———. 1990. "Augustinisme et cartésianisme." In *L'anthropologie cartésienne,* 101–25. Paris: P.U.F.

Roever, N. de. 1893. "Namen van de Geestelijke Personen 1638 die ten gevolge van het placaat van de Staten Generaal van 1622, zich als zodanig hebben bekend gemaakt." *Bijdragen voor de geschiedenis van het Bisdom Haarlem* 18:48–69.

Rogier, L. J. 1947. *Geschiedenis van het Katholicisme in Noord-Nederland in de 16e en de 17e eeuw.* 3 vols. Amsterdam: Urbi et Orbi.

Rose, H. J. 1974. *Gods and Heroes of the Greeks.* London: New English Library.

Rubidge, Bradley. 1990. "Descartes's *Meditations* and Devotional Meditations." *Journal of the History of Ideas* 28:27–49.

Ruler, J. A. van. 1991. "New Philosophy to Old Standards: Voetius' Vindication of Divine Concurrence and Secondary Causality." *Nederlands archief voor Kerkgeschiedenis / Dutch Review of Church History* 71:58–91.

Sainte-Beuve, C.-A. 1930. *Port-Royal.* 7 vols. 9th ed. Paris: Hachette.

Schüling, Hermann. 1969. *Die Geschichte der axiomatischen Methode im 16. und beginnenden 17. Jahrhundert.* Hildesheim: G. Olms.

Schuster, John. 1980. "Descartes' *Mathesis universalis:* 1619–1628." In *Descartes: Philosophy, Mathematics, and Physics,* edited by Stephen Gaukroger, 41–96. Brighton, Sussex: Harvester.

Scotus, John Duns. 1988. *Sur la connaissance de Dieu et l'univocité de l'étant.* Paris: P.U.F.

Sédillot, M. L. Am. 1869. *Les professeurs de mathématiques et de physique générale au Collège de France.* Rome: Imprimerie de Sciences Mathématiques et Physiques.

Senofonte, Ciro. 1989. *Ragione moderna e teologia: L'uomo di Arnauld.* Naples: Guida Editori.

Sepper, Dennis. 1988. "Hobbes, Descartes, and Imagination." *Monist* 71:538–39.

Shea, William R. 1991. *The Magic of Numbers and Motion: The Scientific Career of René Descartes.* Canton, Mass.: Science History Publications.

Sleigh, R. C., Jr. 1990. *Leibniz and Arnauld: A Commentary on Their Correspondence.* New Haven, Conn.: Yale University Press.

Smit, W. A. P. 1928. "De dichter Revius." Ph.D. diss., University of Utrecht.

Sommervogel, Carlos, et al. 1890–1932. *Bibliothèque de la Compagnie de Jésus.* 11 vols. Brussels: Alphonse Picard.

Soprani, Anne. 1987. *Les rois et leurs astrologues.* Paris: MA Editions.

Sorell, Tom. 1993. "Hobbes without Doubt." *History of Philosophy Quarterly* 10, no. 2:121–36.

Stohrer, Walter John. 1979. "Descartes and Ignatius Loyola: La Flèche and Manresa Revisited." *Journal of the History of Ideas* 17:11–27.

Suárez, Francisco. 1856–78. *Opera omnia.* 28 vols. Paris: Vives.

——. 1982. Suárez on Individuation: *Metaphysical Disputation V, Individual Unity and Its Principles.* Edited and translated by J. J. E. Gracia. Milwaukee: Marquette University Press.

Taylor, Richard. 1967. "Voluntarism." In *The Encyclopedia of Philosophy,* edited by Paul Edwards, 8:270–72. New York: Macmillan.

Thomas Aquinas. 1957. *Summa contra Gentiles.* Translated by Charles J. O'Neil. 4 vols. Notre Dame, Ind.: University of Notre Dame Press.

Torre, Jacobus de la. 1882–84. "Relatio seu descriptio status religionis catholicae in Hollandia." *Archief voor de geschiedenis van het aartsbisdom Utrecht* 10:95–240; 11:57–211; 12:189–213, 414–33.

Trosee, J. A. G. C. 1894. *Het verraad van George van Lalain.* 's-Hertogenbosch.

Tuck, Richard. 1988. "Hobbes and Descartes." In *Perspectives on Thomas Hobbes,* edited by G. A. J. Rogers and A. Ryan, 11–42. Oxford: Oxford University Press.

———. 1989. *Hobbes.* Oxford: Oxford University Press.

Verbeek, Theo. 1988. Introduction to *La querelle d'Utrecht.* By René Descartes and Martin Schoock. Paris: Les Impressions Nouvelles.

———. 1992. *Descartes and the Dutch.* Carbondale: Southern Illinois University Press.

Verga, Leonardo. 1972. *Il pensiero filosofico e scientifico di Antonio Arnauld.* 2 vols. Milan: Vita e Pensiero.

Vie de Maistre Jean Baptiste Morin . . . , La. 1660. Paris: Chez Iean Henault.

Vignaux, Paul. 1934. *Justification et prédestination au XIVe siècle. Duns Scot, Pierre d'Auriol, Guillaume d'Occam, Grégoire de Rimini.* Sciences Réligieuses 48. Paris: Bibliothèque de l'Ecole des Hautes Etudes.

Vigne, Guillelmus de la. 1666. Student thesis, Collège du Mont. Cean MS 468.

Vincent, I. 1677. *Discussio peripatetica in qua philosophiae cartesianae principia.* Toulouse.

Visser, Jan. 1966. *Rovenius und seine Werke.* Assen: Van Gorcum.

Vregt, J. F. 1880. "De vroegere Collegiën of Seminariën tot opleiding van Geestelijken voor de Hollandsche Missie." *Bijdragen voor de geschiedenis van het Bisdom Haarlem* 8:1–55, 256–319, 337–420.

Waard, Cornelius de. 1925. "Les objections de Pierre Petit contre le *Discours* et les *Essais* de Descartes." *Revue de Métaphysique et de Morale* 32:53–89.

Wallace, William A. 1981. "Aristotle and Galileo: The Uses of *Hypothesis* (Suppositio) in Scientific Reasoning." In *Studies in Aristotle,* edited by Domenic J. O'Meara, 47–77. Washington, D.C.: Catholic University of America Press.

———. 1992. *Galileo's Logic of Discovery and Proof: The Background, Content, and Use of His Appropriated Treatises on Aristotle's "Posterior Analytics."* Dordrecht: Kluwer.

Watson, Richard A. 1966. *The Downfall of Cartesianism, 1673–1712.* The Hague: Martinus Nijhoff.

Wells, Norman J. 1961. "Descartes and the Scholastics Briefly Revisited." *New Scholasticism* 35:172–90.

———. 1982. "Descartes' Uncreated Eternal Truths." *New Scholasticism* 56:185–99.

Wiggers, Johannes. 1639. *Commentaria de jure et justitia caeterisque virtutibus cardinalibus.* Louvain.

———. 1641. *In primam partem D. Thomae Aquinatis Commentaria. De Deo Uno et Trino.* Louvain.

William of Ockham. 1957. *Philosophical Writings.* Edited and translated by Philotheus Boehner. Edinburgh: Nelson.

———. 1988. *Somme de logique.* Translated by Joël Biard. Mauvezin: T.E.R.

Wilson, Margaret. 1990. "Descartes on the Representationality of Sensation." In *Central Themes in Early Modern Philosophy,* edited by J. Cover and M. Kulstad, 1–22. Indianapolis: Hackett.

———. 1992. "History of Philosophy in Philosophy Today; and The Case of the Sensible Qualities." *Philosophical Review* 101, no. 1:191–243.

Zarka, 1984. "Espace et représentation chez Hobbes." In *Recherches sur le XVIIe siècle* 7:159–80. Paris: CNRS.

———. 1987. *La décision métaphysique de Hobbes.* Paris: Vrin.

CONTRIBUTORS

Roger Ariew
Department of Philosophy
Virginia Polytechnic Institute and
State University
Blacksburg, Virginia

Jean-Robert Armogathe
Ecole Pratique des Hautes Etudes
5ème Section, Sorbonne
Paris, France

Vincent Carraud
Institut de Philosophie
Université de Caen
Caen, France

Edwin Curley
Department of Philosophy
University of Michigan
Ann Arbor, Michigan

Peter Dear
Department of History
Cornell University
Ithaca, New York

Daniel Garber
Department of Philosophy
University of Chicago
Chicago, Illinois

Marjorie Grene
Department of Philosophy
Virginia Polytechnic Institute and
State University
Blacksburg, Virginia

Thomas M. Lennon
Department of Philosophy
University of Western Ontario
London, Ontario, Canada

Jean-Luc Marion
Départment de Philosophie
Université de Paris X, Nanterre
Nanterre, France

Stephen Menn
Department of Philosophy
McGill University
Montreal, Quebec, Canada

Steven Nadler
Department of Philosophy
University of Wisconsin–Madison
Madison, Wisconsin

Margaret J. Osler
Department of History
University of Calgary
Calgary, Alberta, Canada

Tom Sorell
Department of Philosophy
University of Essex
Wivenhoe Park
Colchester, United Kingdom

Theo Verbeek
Department of Philosophy
Rijksuniversiteit Utrecht
Utrecht, The Netherlands

INDEX